THE CITY

POWER AND PARTY IN AN ENGLISH CITY

POLITICS OF THE CITY

POWER AND PARTY IN AN ENGLISH CITY

An Account of Single-Party Rule

DAVID G. GREEN

Routledge
Taylor & Francis Group

LONDON AND NEW YORK

First published in 1981

This edition published in 2007
Routledge
2 Park Square, Milton Park, Abingdon, Oxon OX14 4RN
711 Third Avenue, New York, NY 10017

Transferred to Digital Printing 2007

Routledge is an imprint of the Taylor & Francis Group, an informa business

First issued in paperback 2013

British Library Cataloguing in Publication Data
A CIP catalogue record for this book
is available from the British Library

Power and Party in an English City

ISBN13: 978-0-415-41741-9 (hardback)
ISBN13: 978-0-415-86033-8 (paperback)
ISBN13: 978-0-415-41318-3 (set)

Routledge Library Editions: The City

POWER AND PARTY IN AN ENGLISH CITY

An Account of Single-Party Rule

DAVID G. GREEN
University of Newcastle upon Tyne

London
GEORGE ALLEN & UNWIN
Boston   Sydney

First published in 1981

GEORGE ALLEN & UNWIN LTD
40 Museum Street, London WC1A 1LU

British Library Cataloguing in Publication Data

Green, David G
    Power and party in an English city. – (The new
    local government series; no. 20).
    1. Newcastle-upon-Tyne – Politics and government
    2. Local government – England – Case studies
    I. Title   II. Series
    352.0428'76   JS3759   80-40738

    ISBN 0-04-352094-4

Typeset in 10 on 11 point Times by Red Lion Setters, London
and printed in Great Britain
by Lowe & Brydone Limited, Thetford, Norfolk.

CONTENTS

TABLES

PREFACE

This book is an account of how decisions are taken by the state at the level of the locality. More specifically, it is an account of the private policy-making activities of a ruling Labour group of councillors in the major English city of Newcastle upon Tyne. The data were acquired more or less equally by means of a questionnaire and by participant observation (the author was a Labour member of the authority being studied).

The decision-making processes of the full Labour group and of the Labour groups on council committees are examined in some detail. Chapters are devoted to the Labour group's relationship with the local party organisation, the role of the group's leadership, party discipline, the problems faced by individual backbench councillors, and to an examination of the openness of the local authority's decision-making machinery.

The last part of the book is devoted to a discussion of aspects of some traditional and modern theories of democracy and specifically to what the author sees as the inadequate advocacy of participatory democracy in recent years. The utility of participation for the efficiency of government is explored in the light of a theory of knowledge which is seen as more appropriate for modern natural and social scientific endeavour. Finally, proposals intended to improve the quality of local authority decision-making are put forward. They are designed to encourage rational free criticism of local authority activities by citizens. To that end, the enactment by Parliament of a set of Principles of Good Government – to be enforced by the ordinary courts of law – is proposed.

I would like to thank my colleagues in the Department of Social Studies at the University of Newcastle upon Tyne for providing an atmosphere congenial to work. Specifically I would like to thank Elizabeth Gittus, Jon Gower Davies and Norman Dennis, who are fellow members of what we have come to call in our more grandiose moments the north-east school of political sociology. I would like to single out Norman Dennis, who acted as supervisor of my postgraduate studies and who has been a marvellous source of inspiration over the last few years. I am also indebted to Professor Hugh Berrington and Peter Jones of the Politics Department at the University of Newcastle upon Tyne, who read and commented on part of the typescript; and to Professor Peter Richards of Southampton University, not only for reading and commenting on an early draft, but also for his encouragement. Needless to say, any inadequacies and/or errors are entirely my own responsibility. My thanks are also due to Margaret Riley, who typed the manuscript and who stoically put up with the constant stream of corrections and modifications. Finally I acknowledge the support received from the Social Science Research Council from November 1973 to November 1976. I am indebted to Alan Norton, editor of *Local Government Studies*, for permission to reproduce copyright material first published in *Local Government Studies*, vol. 6, no. 1 (1980).

To
My Mother and to Catherine

Chapter 1

INTRODUCTION

> Throughout the world, in the name of progress, men who call themselves communists, socialists, fascists, nationalists, progressives, and even liberals, are unanimous in holding that government with its instruments of coercion must, by commanding the people how they shall live, direct the course of civilization and fix the shape of things to come ... the premises of authoritarian collectivism have become the working beliefs, the self-evident assumptions, the unquestioned axioms, not only of all the revolutionary regimes, but of nearly every effort which lays claim to being enlightened, humane, and progressive.
> (Walter Lippman, 1937, pp. 3–4)

Any account of an important aspect of reality must be selective unless it is to be swamped by a mass of irrelevant details. What is important is that the factual beliefs and personal preferences which have determined the particular choice of relevant research material and from which any critical or analytical comments emerge should be made explicit. Hence the next few pages contain an account of the factors which have influenced this interpretation of local government in Newcastle upon Tyne.

Empirically the study is an attempt to contribute to the task of filling a gap in the now growing body of literature on the government of our towns and cities. Local government, in the overwhelming majority of towns and cities in this country, has for many years been one-party government (Maud Committee, 1967, Vol. 1, pp. 6–8 and 106–15) with ruling groups of councillors typically deciding in private what policies they will support. Newcastle upon Tyne was no exception. To paint an accurate picture of the way in which the local authority utilised its powers it was therefore necessary to study not only the authority's official decision-making apparatus but also the internal decision-making machinery of the majority party. The latter field of study is the principal area which has so far been neglected.

No doubt one of the main reasons for the limited extent of the work in this field is that it is very difficult to obtain access to the internal decision-making processes of controlling groups. The present author has been able to overcome this problem by becoming a member of a controlling group of councillors: the Labour group in Newcastle upon Tyne. By acting as a participant observer I have been able to experience directly how Labour councillors conducted themselves.

Although membership of the Labour group of councillors provided information which would otherwise not have been available it also had a disadvantage, namely, that similar access to the private activities of the minority Conservative group of councillors was thereby ruled out. Clearly it was not possible to be an insider to both Labour and opposition councillors at once. This might be thought to be a serious limitation but because Newcastle City Council was run on very strict party lines members of opposition parties only had such influence as the Labour group allowed them. In practice, opposition participation was negligible and its influence slight. This meant that few, if any, insights of importance into how decisions were arrived at could have been obtained by having access to the Conservative party's internal decision-making activities.

Participant observation was not the only method of research employed. Documentary sources were also used where appropriate and, more important, in the autumn of 1977 forty of the forty-four Labour councillors were interviewed by means of a questionnaire (see Appendix). The findings yielded by the questionnaire and by the author's direct observations more or less equally provide the basis for this account of local government in Newcastle upon Tyne.

Two main linked concerns lie behind the author's approach to carrying out the study. I have attempted to summarise these concerns below before dealing with them in more detail.

A major factor has been a growing sense of dissatisfaction with the disappointing results brought about by the expansion of state power since the Second World War. At the same time, however, I have remained sympathetic to the collective provision of goods and services where this is the best way to cater for people's needs and I have therefore been unsympathetic towards contemporary theories which are in essence simply hostile to all forms of collectivism. To accommodate these somewhat conflicting concerns I have attempted to adopt a critical stance towards state activity which is essentially favourable to collective provision where it is the best way to provide for popular needs and yet at the same time suspicious of the state as a potential menace not only to individual liberty but also to individual well-being.

The second factor which has determined my approach to the study has been a sense of dissatisfaction with many modern theories of democracy. John Stuart Mill advanced a developmental theory of democracy which stressed the value of participation as a means to self-improvement and the necessity for the machinery of government to allow men of intellectual and moral worth in the community to exercise a continuing influence on the conduct of government. Mill's theory gained wide currency in England and America and enjoyed predominance until it began to be replaced from the Second World War onwards. It lost ground to a theory resting on belief in the overriding importance of recognising the realities of government as they were then being experienced. This theory focused on the competition between elites for power by means of elections. Supporters of this 'pluralist' perspective were in their turn attacked for their undue emphasis on mere description. They were accused of being too willing to accept men as they were. The antipluralists advocated participatory theories of democracy, often returning to Mill and also to Rousseau for their inspiration. Participatory theories of democracy are now widely supported but they have been susceptible to the charge, advanced by democratic elitists, that it is unrealistic to advocate a more participatory style of government because it would be inefficient. Advocates of participatory theories have tended to share this presumption and laid stress on participation as a means of self-improvement and development rather than on its value for efficiency. I will attempt to argue below that participation (of the kind the present author wishes to advocate) can be justified not only on ethical grounds but also on grounds of efficiency, that is, that better government will result.

THE HISTORICAL CONTEXT

C. Wright Mills stressed the importance of studying human activity in its wider social and historical context. He wrote:

> Try to understand man not as an isolated fragment . . . Try to understand men and women as historical and social actors . . . Before you are through with any piece of work . . . orient it to the central and continuing task of understanding the structure and the drift, the shaping and the meanings, of your own period, the terrible and magnificent world of human society in the second half of the twentieth century.
> (Mills, 1970, pp. 247–8)

What are the salient features of this 'terrible and magnificent world' and how is this study related to them?

During the sixteenth and seventeenth centuries England began to develop into a market economy: 'More than a century before Adam Smith published *The Wealth of Nations* men were ready to accept 'self interest and economic freedom as the natural basis of human society' (Lipson, 1947, p. lxvii). The old society based on custom and status took time to die but by the nineteenth century England had become a fully-fledged market economy (Thompson, 1968, p. 73). Freedom of choice had become the dominant principle:

Individuals were free to choose their religion, their pattern of life, their marriage partners, their occupations. They were free to make the best arrangements they could, the best bargain they could, in everything that affected their living.
(Macpherson, 1966, p. 6)

This transformation led to a change in the form of government. By revolutionary action in the seventeenth century arbitrary government – 'England's relatively weak version of royal absolutism' (Barrington Moore, 1973, p. 4) – was ended and the liberal state inaugurated. During the eighteenth century a liberal system of government was firmly established. Its essential characteristics were these. Governmental power was in the hands of individuals subject to periodic elections at which there was a choice between candidates and parties. The electoral system was underpinned by the existence of political freedoms which made it possible to organise effectively against the powers that be. The state recognised fields of personal and collective behaviour which were beyond the province of the state. The government provided and enforced the legal and other conditions necessary for the free market economy to function. The electorate was not democratic and the society not equal. The interests served were largely those of the commercially minded landed upper class.

However within the logic of the system there were no defensible grounds for withholding the vote from other sections of society (Macpherson, 1966). Liberal society was justified as a form of social organisation which provided equal individual rights and equality of opportunity. Freedom of association, freedom of speech and publication had been demanded in principle for everybody. These freedoms were used by the disenfranchised majority to demand the vote. It took many decades but ultimately the vote was achieved for all. At about the same time as the mass of working men were winning the right to vote, and partly as a result of their success, a new transformation began to take place.

Much pre-Victorian libertarian writing was concerned with controlling the state because, with its monopoly on the use of coercion, it was seen as the principal threat to individual freedom. But during the nineteenth century writers and activists who placed individual freedom first found themselves forced more and more into socialistic modes of thinking as a result of what the Webbs described as 'the devastating torrent of public nuisances' produced by the Industrial Revolution (Webb and Webb, 1963, p. 72). As Hobhouse pointed out, 'liberty without equality is a name of noble sound and squalid result' (Hobhouse, 1911, p. 86). Perhaps no other leading figure of that period exemplifies this transformation better than J. S. Mill. He began his life as a Benthamite Liberal but, as he records in his autobiography, during his later life he classed himself 'decidedly under the general designation of socialists' (Mill, 1924, p. 196). Such thinking led individuals who were concerned to protect and promote individual freedom to concentrate their efforts on extending the power of the state to control the activities of private persons and organisations which threatened the individual liberty of others, and on advocating increased direct provision by the state of goods and services not catered for by market forces. Dicey – writing in 1905 – described the period from 1865 to 1900 as the 'period of collectivism'.[1] By the end of the nineteenth century, in his view, the collectivist approach was firmly established (Dicey, 1905, pp. 62–9 and pp. 258–301).

Since then collectivist attitudes have continued to enjoy preeminence and the level of state activity has grown steadily. By the 1950s Britain had a fully-fledged welfare and regulatory state. Peacock and Wiseman have undertaken the complex task of measuring this growth in public expenditure since the end of the last century. Total government expenditure grew, as a proportion of the gross national product, from just under 8 per cent in 1890 to about 37 per cent in 1955 (Peacock and Wiseman, 1967, p. 42). In recent years public expenditure, as a proportion of gross domestic product, has risen sharply to a peak in the mid-1970s. An OECD study shows that general government expenditure in the United Kingdom, as a proportion of gross domestic product, increased from 34 per cent in 1962 to 44 per cent in 1975 (OECD, 1978, p. 16). The expenditure of local authorities is a significant element in total public expenditure. In 1951 the current expenditure of local government disposed of about seven per cent of the gross domestic product. In 1961 the figure was about 9 per cent. Since 1961 the proportion of gross domestic product being disposed of by local authorities has risen rapidly from 9 per cent to 19 per cent in 1974. An examination of employment statistics reveals a similar picture. An increasing number of people now find employment in the public sector. By 1975 this figure had reached about 32 per cent of the total workforce. Of public sector employees, about 40 per cent are employed by local authorities.

Thus the arbitrary state gave way to the liberal state as a result of pressures generated by the rising competitive market society. This liberal state conceded the democratic franchise; and the liberal-democratic state was transformed into the modern welfare and regulatory state.

Whether or not state power will continue to grow remains to be seen. At the time of writing the Conservative government elected in May 1979 was showing signs of controlling the growth of state activity and of reducing the proportion of gross domestic product taken up by public expenditure. But even if they implement the most radical measures they have proposed so far the state will remain a dominant force in contemporary Britain.

Just as the state is a pervasive force in contemporary Britain, so the local state in Newcastle upon Tyne is a pervasive force on Tyneside. Newcastle upon Tyne is the regional capital of the north east of England. In 1979 the local authority was by a long way the largest single employer in the area, employing about 18,000 people across a wide range of occupations. Of the 110,000 dwellings in Newcastle about half were owned by the authority. With its control of housing, schools, the personal social services and recreation facilities, its planning powers and more, the authority's impact on the lives of Newcastle's population of just under 300,000 was tremendous.

Most socialists have regarded growing state power as relatively unproblematic. Indeed, many regarded criticism of abuses of state power as disloyal, a tendency attacked by Richard Crossman, writing in 1956:

> Far too many socialists regard it as reactionary (or at least as no part of a socialist's duty) to take up the cudgels for the individual citizen who feels that his rights have been violated by a Department of State, a public board or a semi-public authority.
> (Crossman, 1956, p. 20)

'Socialism', Crossman asserted, 'must challenge power which is either irresponsible or only semi-responsible – in whatever hands that power rests' (p. 7). Bertrand Russell would have agreed with him. Four years earlier he had warned socialists that, unless action were taken to prevent it, the former power of capitalists which they had fought to control might be simply taken over by officials:

> This tyranny of officials is one of the worst results of increasing organisation, and one against which it is of the utmost importance to find safeguards if a scientific society is not to be intolerable to all but an insolent aristocracy of Jacks-in-office . . . The increased power of officials . . . has the drawback that it is apt to be irresponsible, behind-the-scenes, power, like that of Emperors' eunuchs and Kings' mistresses in former times . . . Liberals protested, successfully, against the power of kings and aristocrats; socialists protested against the power of capitalists. But unless the power of officials can be kept within bounds, socialism will mean little more than the substitution of one set of masters for another: all the former power of the capitalist will be inherited by the official.
> (Russell, 1976, pp. 46–7)

The view that the power of the state is unproblematic is no longer tenable by anyone concerned to enhance and protect individual freedom. Certainly significant threats to the freedom of the individual remain outside the public sector but the state itself must also be regarded as a menace of at least comparable magnitude. This point of view is beginning to find wider acceptance on the left of the Labour Party. Judith Hart, in an article in *Tribune*, wrote: 'We have witnessed, and indeed created, in our lifetime the growth of a state . . . apparatus which exercises great economic and social power over the lives of individuals.' She describes this apparatus as 'formidable', and ranks it along with the power of multinational corporations as one of two major threats to individual freedom in modern Britain (Hart, 1977).

If state power is accepted as potentially problematic, at least two major questions are raised. First, is state activity problematic by virtue of its scope and, if so, should it be reduced? The issue of the proper scope of government activity is of great importance, but it is not the subject of this book. Secondly, given the scope of state power, and given the inevitability of a high level of state activity for the forseeable future, how well is this power being exercised? In carrying out this study of the state at the level of the locality I have concentrated on this second question. Later in this introduction I have set out the factors which influenced the way in which I went about answering it. Before doing so I have reviewed the intellectual context in which the study is set.

THE INTELLECTUAL BACKGROUND

It was not until the early nineteenth century that significant theorists began to accept that 'one man, one vote' need not be a danger to property (Macpherson, 1977, p. 10). Bentham and James Mill were the first systematic thinkers to take this view, although they did so with reservations. They saw society as a collection of individuals constantly seeking to maximise their own wealth and power at the expense of others. Social good was regarded as the greatest happiness of the greatest number and the rightness of actions depended on their contribution to the attainment of the greatest happiness of the greatest number. Happiness was regarded as the amount of individual pleasure minus pain. They wanted a political system which would produce governments able to protect and promote the competitive free market society which they favoured and which also enabled citizens to protect themselves from tyrannical acts committed by the government. Bentham assumed that governments were prone to 'depredation and oppression' (Bowring, ed., 1962, Vol. 9, p. 47) and argued that the only way to prevent rulers from pursuing their own selfish interests was to make them removable by the very people who might suffer at their hands. By 1820 this reasoning led Bentham to advocate universal manhood suffrage (Bowring, ed., 1962, Vol. 3, p. 599).

Earlier theories of democracy had contained a prescriptive element; they looked for new attitudes and behaviour in men. The protective democracy of James Mill and Bentham did not, but the democracy advocated by John Stuart Mill was strongly prescriptive. For J. S. Mill all government which aimed at being good was 'an organisation of some part of the good qualities existing in the individual members of the community for the conduct of its collective affairs' (1972, p. 195). One criterion therefore of the goodness of government was the extent to which 'it tends to increase the sum of good qualities in the governed'. The second criterion was the degree to which the machinery of government was adapted to take advantage of good qualities – moral intellectual and active – in the community and to 'make them instrumental to the right purposes' (p. 193). J. S. Mill's vision of democracy was adopted by many twentieth-century intellectuals, notably A. D. Lindsay, Ernest Barker and R. M. MacIver. It was, in Macpherson's words, 'the democracy that World War I was to make the world safe for' (1977, p. 48). It enjoyed pre-eminence in England and America until the Second World War.

From the Second World War onwards a major preoccupation of students of politics was with the inadequacy of what they saw as 'classical democratic theory'. They attempted to reformulate democratic theory so that it took account of the realities of government in America (and to a lesser extent England) – realities which they felt 'classical theory' ignored. They found classical theory inadequate principally because it assumed a high level of citizen participation in government, when in fact citizen involvement was low. A second preoccupation among political scientists at this time was with the threat of totalitarianism. In the 1920s and 1930s Europe had witnessed the collapse of democratic regimes and their replacement by totalitarianism in its fascist form. Many academics in the postwar period were determined to avoid a recurrence of these events. This determination was reinforced by what was seen as the growing threat presented by Soviet totalitarianism. What were the theories of democracy advanced by writers during this postwar period?

A convenient starting point is the work of Joseph Schumpeter. Writing first in 1942 he was concerned with what he saw as defects in the classical doctrine of democracy. His main criticism was that it assumed a high level of participation and decision-making by citizens which did not in reality take place. In his influential *Capitalism, Socialism and Democracy* he argued that the distinctive feature of democracy was competition for leadership rather than direct citizen involvement. He offered a revised formulation of the democratic method. It was defined as 'that institutional arrangement for arriving at political decisions in which individuals acquire the power to decide by means of a competitive struggle for the people's vote' (1947, p. 269).

Later writers went further than Schumpeter in their acceptance of low citizen participation. They argued that it was desirable to the extent that it was functional for stability. This point of view was advanced by Berelson, Lazarsfeld and McPhee in their influential work, *Voting* (1954). They contrasted the behaviour of the American voter with the model of the citizen postulated, as they saw it, in classical theory. They found that the average citizen failed to display the expected qualities. But although they observed deficiencies in individual voters they noted that the system of democracy as a whole survived and worked. Exactly how it survived was a little mysterious, or in their own words: 'Where the rational citizen seems to abdicate, nevertheless angels seem to tread' (1954, p. 311). Other theorists went still further. They regarded high participation as a possible threat to liberal democracy (Milbrath, 1965, pp. 142–55; Lipset, 1960, pp. 32–3).

For theorists in the tradition described above the election was the mechanism which made government democratic. Another group of theorists attached less importance to elections; they argued that competition between groups in the inter-election period was more important. In 1951 David Truman advanced a group theory of politics based on the earlier work of Arthur Bentley. Truman argued that there was an increasing tendency for groups to arise to promote or protect the interests of their members. The formulation of public policy was explained as the outcome of conflict between groups which he valued both as guarantors of individual liberty against the power of the state and as means by which individual demands could be effectively articulated. Groups also offered an opportunity for individual participation.

What these postwar analyses had in common was their recognition of the existence of elites in American society, combined with the claim that these elites were in competition with each other and that it was this competition which gave American government its democratic character. This analysis, however, began to be countered in 1953 with Floyd Hunter's *Community Power Structure*, a study of Atlanta from which he concluded that Georgia's capital was dominated politically by a business elite. He found more than one elite – in addition to a handful of leaders there were several cliques – but he did not find the competition between elites which Schumpeter and his followers thought so crucial. Also influential was C. Wright Mills's *The Power Elite*, originally published in 1956. He argued that behind the democratic facade lay a social and economic elite which really ran things.

These analyses were countered by Robert Dahl and his associates Nelson Polsby and Raymond Wolfinger. From 1957 to 1959 they carried out an extensive study of New Haven. Their theory of democracy came to be called the pluralist theory of democracy (though Dahl preferred to speak of polyarchy). Dahl found that citizens acted through pressure group activity and also through the electoral system. He found that citizens were effective through elections and that power was dispersed in a non-cumulative fashion amongst interest groups

(1961). He reported that power in New Haven was dispersed amongst a number of issue-oriented elites. He recognised that most citizens were politically inactive – the apolitical stratum – but argued that their indirect collective influence through the electoral system was high. He also recognised that the complexity of the political process and the variety of interests to be considered gave leaders considerable freedom of action. External constraints on leaders were few but in spite of this he found that in practice they showed restraint because they adhered to what he called the 'democratic creed'.

In two articles published in 1962 and 1963 the work of Dahl and his associates was criticised by Bachrach and Baratz. (They expanded on their view in a book published later in 1970.) Their criticisms were largely methodological. Dahl's method involved isolating particular issue areas and studying how actual decisions were reached in those areas. In this way a body of data about what actually happened was provided. Bachrach and Baratz criticised Dahl's emphasis on studying decisions and followed Schattschneider in asserting that what was important was the structural setting in which policies were arrived at. He had argued that 'all forms of political organisation have a bias in favour of the exploitation of some kinds of conflict and the suppression of others because *organisation is the mobilisation of bias*. Some issues are organised into politics while others are organised out' (1960, p. 71, his italics). For Bachrach and Baratz the mobilisation of bias could lead to what they called a non-decision, which they defined as 'a decision which results in suppression or thwarting of a latent or manifest challenge to the values or interests of the decision maker' (1970, p. 44). Such non-decisions they felt were overlooked by the issue method of Dahl and his associates, which focused on concrete decisions. Leaders, they argued, might well be able to ignore citizens' interests and be able to alter, exploit or suppress conflict within the system. This was what Bachrach and Baratz found in Baltimore, where they carried out their research. It was a city in which 40 per cent of the population – the blacks – were excluded from playing an effective part in government by the white-dominated establishment (1970, p. 53).

The pluralist perspective was subjected to more than methodological criticisms. Dahl and his colleagues were strongly criticised for being too willing to accept people as they found them. Democratic theory, it was argued, ought to be prescriptive and not merely descriptive. David Ricci is representative of this sort of thinking. He writes:

> It ... is a major contention of antipluralist scholars that we cannot accept the present behaviour of men as final proof that they are capable of no better ... We must therefore go beyond the evidence in order to plan in terms of what men *ought* to do, *ought* to want, *ought* to receive, *ought* to pay for. (1971, pp. 210-11, his italics)

The antipluralists mainly thought that men *ought* to want more participation. Western democracy was not democratic enough. Government ought to be made more democratic by extending participation. Two main demands may be distinguished. First, there were demands for more participation in industry. Secondly, there were demands for more participation in government. The latter demands involved the sort of decentralisation that would allow more direct democracy, particularly based on the neighbourhood; and demands for greater consultation and participation in decision-making. In England more participation was demanded in local authority decision-making not only in planning physical changes but also in council house management and in the running of schools through parent or parent-teacher associations.

In 1963 Duncan and Lukes criticised contemporary democratic theorists for downgrading the place of participation in modern democratic theory. They stressed that participation developed character. In a criticism of pluralism – which he called democratic elitism – published in 1967 Bachrach argued that it was both empirically unsound and normatively inadequate. He asserted the need for a new theory of democracy: 'one that is founded on the self-developmental objective and one that at the same time firmly confronts the elite-mass structure characteristic of modern societies' (1969, p. 99). Democracy, he urged, should have as its overriding objective the self-development of the individual. He very strongly advocated the necessity for industrial democracy, on the ground that it was more likely to attract the interest of modern-day citizens. Similar participatory theories of democracy continue to be advanced (for example, Pateman, 1970; Macpherson, 1977). In many quarters participatory theories enjoy pre-eminence.

For the present I am content to review the position to date. I return to my contention that modern advocates of participatory theories have laid too much stress on ethical arguments for participation and devoted insufficient attention to its practical advantages in the concluding chapters. I shall attempt to comment on this problem in the light of the findings of this study of local politics in Newcastle upon Tyne.

THE SELECTION OF RESEARCH MATERIAL

In considering the decision-making machinery of the local authority as a whole I have been strongly influenced by the work of John Stuart Mill.

> The first question in respect to any political institutions is how far they tend to foster in the members of the community the various desirable qualities . . . moral, intellectual and active. The government which does this best has every likelihood of being the best in all other respects, since it is on these qualities, so far as they exist in the people, that all possibility of goodness in the practical operations of the government depends. (1972, p. 193)

For Mill this was one of the two main criteria of the goodness of a government. The second was the quality of the machinery of government itself, that is, 'the degree in which it is adapted to take advantage of the amount of good qualities which may at any time exist and make them instrumental to the right purposes' (ibid.). Mill's two criteria — first, the degree to which the government increases the sum of good qualities in the population and, secondly, the extent to which the machinery of government is adapted to take advantage of these good qualities — are no less relevant today than they were in the middle of the last century.

In discussing the Labour Party's policy-making machinery Mill's two criteria were again very much to the fore. I have focused particularly on Mill's second criterion, and specifically on the degree to which the party's machinery took advantage of the good qualities among one part of the community — the elected representatives of the people. If a problem existed, then I have asked to what extent was a councillor with the appropriate knowledge, or who displayed interest, or who was willing to put in some effort, able to bring his knowledge, interest or effort to bear on finding a solution? What machinery existed for deliberating about the appropriateness of goals and the best means of achieving them, and for scrutinising the implementation of policies? Could councillors with moral, intellectual and active qualities bring them to bear on these activities? Or, on the contrary, did the machinery of government tend to exclude such influences (deliberately or otherwise)?

A second set of concerns has been with those issues most closely identified with the name of Robert Michels. As a result of his studies of the German Social Democratic Party Michels formulated what he described as the fundamental sociological law of political parties: 'It is organisation which gives birth to the dominion of the elected over the electors, of the mandatories over the mandators, of the delegates over the delegators. Who says organisation says oligarchy' (Michels, 1968, p. 365). For Michels, writing first in 1911, the principal cause of oligarchy in ostensibly democratic political parties was the 'technical indispensability of leadership' (p. 364). There were other factors. For example, there was the need for leadership felt by the mass and their sense of gratitude towards party leaders. There was the tendency for leaders to regard themselves as indispensable — *le parti c'est moi* — and to regard criticism of themselves as an act of treason against the organisation, thus justifying the use of repressive methods against their opponents (pp. 220–3). But, important though these other factors were, for Michels the essential factor was that oligarchy resulted automatically from the need to organise on a large scale: it was 'the outcome of organic necessity' (p. 366). Michels's concerns remain important today and in this analysis of the activities of the Labour Party in Newcastle I have attempted to employ the insights he has provided.

Thus, in choosing relevant research material the questions raised by Mill and Michels have been important. The following speculation has been equally important in determining my approach: the best guarantee of efficient government is that its machinery should facilitate and encourage free criticism of its own activities. By efficient government I mean government that is more likely to carry out policies that are (a) based on better information and (b) in the interest of, or at least acceptable to, a wide rather than a narrow range of individuals or groups. In referring to better information I am concerned with two main issues. I am concerned first with the problem of what statements may correctly be regarded as true. Here I contend that free criticism (by free I mean that all who wish to contribute may do so on equal terms) makes it more likely that the correct version of the facts will be arrived at; conversely, in the absence of such an approach it is far more likely that deliberators will fall into error. Secondly, I am concerned with the issue of relevance, that is, with which facts and which values come to be regarded as relevant to a problem, and to finding a solution to it. I contend that free criticism makes it more likely that all the relevant facts and all the relevant values will be brought to light. Social problems, in relation to which there is a single obvious value to be uncontroversially applied, are extremely rare; almost always there is more than one value that could be applied. Such values may result from some specific sectional interest or from some attitude towards what the public good is or should be (such values may be selfish or altruistic). If only a narrow circle of deliberators is involved, it is more likely that they will have themselves adopted a narrow set of values. Similarly, they are more likely to be unaware of all the values regarded as relevant by other actors. This, in turn, may affect their definition of the situation and influence their attitude towards the relevance (or the truth) of some facts and statements.

The second element of efficiency, as defined above, raises the issue of whether or not a wide or narrow range of interests or values are being served. My contention is that judgements of value are more likely to be made in a way that results in 'workable' or 'satisfactory' outcomes if they are not made by individuals or small groups acting alone but rather by a number of individuals or groups in free and open competition with each other.

For the present I simply assert that these are the problems I regard as important and the theses I regard as worthy of investigation. I attempt to justify my point of view towards the end of the book, in Part Six. Also in this final section I discuss whether or not any lessons may be learnt from the study for the application in practice of democratic principles under modern conditions.

The study covers the period from October 1973 to October 1979. The

Labour Party was in control of the council for most of this time, from 1 April 1974 onwards. (It had won control of the council in the elections of May 1973, but because of local government reorganisation there was a transition year.) Throughout the study period I was an active member of the local Labour party. For parts of the period before December 1975 I attended Labour group meetings either as a party delegate or as an electoral candidate. From December 1975 and throughout the remaining period of the research I was also a member of the Labour group of city councillors. During this time I served on the following committees: the Social Services Committee, the Housing Committee, the Housing Management Committee (as vice-chairman from May 1978), the Direct Labour Organisation Committee, the Development, Planning and Highways Committee, the Economic Development Committee, the Blakelaw Area Housing Management Committee (as chairman from October 1977) and several subcommittees.

NOTE: CHAPTER 1

1   Dicey distinguishes between three main currents of legislative opinion which over-lapped each other: the period of old Toryism or legislative quiescence (1800–30); the period of Benthamism or Individualism (1825–70); and the period of Collectivism (1865–1900).

Part One

THE LOCAL PARTY

Chapter 2

PARTY MEMBERSHIP AND ORGANISATION

THE BACKGROUND

The existence of organised groups of councillors on local councils is not a recent phenomenon. Elections were fought on party lines in many towns from the very moment the 1835 Municipal Corporations Act became law (Redlich and Hirst, 1903, Vol. 1, pp. 264–5). In Leeds there were contests between party groups for control of the Poor Law Guardians as well as for the council. The Exeter City Council 'was firmly divided on political lines from the outset' (Newton, 1968, p. 302). Later in the century the famous Liberal caucus at Birmingham was established. The Birmingham Liberal Association had been founded in 1865. It was given great impetus by the 1867 Reform Act which conferred the right to vote on about 30,000 individuals in that city. The Liberal Association was determined to secure their support and in order to do so set up an organisation in each ward. They captured both the city council and the school board in 1873 and Joseph Chamberlain was made lord mayor (Ostrogorski, 1902, Vol. 1, pp. 161–5). According to Briggs, 'city government was never quite the same again' (Briggs, 1963, p. 195).

It was not until the 1880s that the first working men stood for election to the Newcastle City Council. A working man had stood for election, unsuccessfully, to the school board in 1871, but no other elections were contested by working men until the 1880s. A major reason was that many working men did not have a vote. The local franchise was enjoyed only by ratepayers and until well into the second half of the century many labourers' cottages were not rated. Until 1880 there was also a large property qualification for council candidates (though not for the school board). In addition to this, most working men tended to be Liberals or Radicals and pursued their interests through the Liberal Party. By the 1880s, however, many Radicals and working men had become disenchanted with the Liberal Party. The Fabian Society and the Social Democratic Federation were both established at this time. The Independent Labour Party was established a little later, in 1893, and the Labour Representation Committee (which changed its name to the Labour Party in 1906) in 1900. From the outset they fought council, school board and other elections.

The first working man to become a councillor in Newcastle was the president of the local Trades and Labour Council. He was elected in 1883 as a Liberal in the labour interest. The first labour candidate to stand for election independent of the other parties was the president of the local branch of the Social Democratic Federation. He stood in 1885, unsuccessfully. Then in 1889 came the first successes for independent labour candidates, three members of the Social Democratic Federation were elected on to the school board. Five working men also stood for the council that year, mainly as a result of Trades Council support, but they all lost. In 1890 four more stood but they also lost. In 1891, the Trades Council secretary, a member of the Amalgamated Shipwrights Society, was elected. In 1892 Arthur Henderson, then an iron moulder, later to become a Labour MP and Labour Party secretary for twenty-two years was elected onto the council as a Liberal in the labour interest. In the 1892 school board elections the three socialists lost, though one was returned again in 1895.

During the late 1890s the Municipal Reform Committee supported labour candidates. In 1897 labour representatives fought six seats, mostly supported by the Municipal Reform Committee. In 1898 and 1900, however, they only fought one seat and in 1899 none. Then in 1901 the Newcastle upon Tyne Labour Representation Committee was established, and members of the Municipal Reform Committee appear to have joined. In 1902 the chairman of the Labour Representation Committee was elected. He was the first Labour councillor, independent of other parties. In 1904 Alexander Wilkie, a national leader of the Amalgamated Shipwrights Society and member of the National Executive Committee of the Labour Representation Committee, was elected onto the council. By the First World War there were seven Labour councillors. Six were working men, one was a merchant. After the First World War the Labour party in Newcastle gradually grew in strength. By 1937 there were twenty-five Labour members: twenty were councillors (out of a total of fifty-seven) and five were aldermen (out of a total of nineteen).

Nevertheless Labour did not take control of the council until after the Second World War. At the 1945 elections thirty-one Labour councillors were elected, along with twenty-six Progressives. Counting the aldermen, Labour had a majority of twelve. They then lost control of the authority in 1949 and did not regain it until 1958. They again lost control in 1967 and were never to recapture control of the old authority. Labour did, however, decisively gain control of the new authority which came into being on 1 April 1974.

PARTY MEMBERSHIP NATIONALLY

During the period of this study there was national concern about declining party membership figures. For example, in 1977 three Labour Party members (two of whom were full-time employees at party headquarters and one of whom had recently been a full-time employee) felt so concerned about the state of the party that they published privately a pamphlet drawing attention to the problem as they saw it and advocating some positive steps to put it right (Clarke, Humphris and James, 1977). In the section entitled 'Party membership: fact or fiction?' they write:

> The real situation . . . shows that we have ceased . . . to be a mass party. This is because the official membership figures are entirely fictional . . . If such a haemorrhage of membership continues, the whole future of the party will be in doubt. The National Executive's first priority must be to re-shape the party organisation to prevent the withering away of the grass roots of the party. (p. 3)

Individual membership figures have been published from 1928 onwards. In that year there were 214,970 individual members. This rose to a peak of 1,014,524 in 1952. By 1976 the figure had declined to 659,058. However, this 1976 figure is in fact a substantial overstatement, as Clarke *et al.* point out. The figure is an aggregation not of the real membership of constituency parties but of the membership figures submitted by constituencies in order to secure affiliation to the national party. In 1952 constituency parties affiliated to the national party on a minimum membership of 240 and figures over 240 could be relied upon to correspond roughly with the actual number of members. In 1976 the minimum affiliation figure was 1,000 members so that all that was known from the official figures was that the actual membership of a constituency party affiliated on the minimum figure was 1,000 or less. Only eighty-six of the constituency parties in 1976 reported more than 1,000 individual members, only twelve more than 2,000 and only five more than 3,000.

If the official figure was fictional then what was the true figure? A number of estimates have been made. The Houghton Committee on Financial Aid to Political Parties, of which the national agent of the Labour Party was a member, estimated that individual membership in 1974 was nearer 300,000. They calculated the average membership of constituency parties in 1974 for the three main political parties. They found that in the Conservative Party the figure was 2,400, in the Liberal Party that it was 300 and in the Labour Party, 500 (Houghton Committee, 1976, p. 31, para. 5.10). In February 1977 the Gallup organisation carried out a survey for the BBC television programme

Panorama. They contacted the agent, secretary or chairman of 315 constituency Labour parties and asked them to estimate the number of individual members in their own organisation. They found that 15 per cent of constituency Labour parties had 250 members or less; that 45 per cent of constituency Labour parties had 500 members or less; and that 86 per cent had 1,000 members or less (Martin and Martin, 1977, p. 463). On the basis of individual party members per 100 party voters – even on official Labour Party figures – Labour was among the smallest socialist parties in Western Europe (Hayter, 1977, p. 14). The figures for youth membership were no more impressive. About half of the constituency parties were without youth branches in the mid-1970s; there were 356 branches and almost 5,000 active members (Labour Party, 1977a, p. 20). It should be mentioned that *total* membership, which included members of affiliated trade unions, socialist societies, co-operative societies and other groups, had not declined to nearly the same extent. The peak year for total membership was 1957 when the figure was 6,582,549. By 1976 the figure was 6,459,127. However, the vast majority of these members were trade unionists who paid the political levy and their commitment to the party was confined to routinely paying their trade union's political levy through their pay packets.

The declining membership created three problems which, in turn, made it still more difficult to halt and reverse the decline. First, it meant poor organisation. Indeed the weakness of the party's organisation had been recognised since the mid-1950s, particularly since the publication of the Wilson Report to the 1955 annual conference (Labour Party, 1955, pp. 63–105). In 1977 Clarke *et al.* said this:

> More than twenty years ago, the committee on party organisation chaired by Harold Wilson made this stark statement ' . . . compared with our opponents we are still at the penny-farthing stage in a jet-propelled era, and our machine, at that, is getting rusty and deteriorating with age'. (p. 2)

The three authors continued: 'We have now entered the age of Concorde but our machine is now an ailing bone-shaker, in danger of collapsing altogether.' A particularly serious problem was the declining number of full-time organisers employed by the party. Clarke *et al.* report that there were 87 full-time agents for 623 constituency Labour parties and that the numbers had fallen by nearly half in seven years, from 160 in 1970 to 87 in 1977. Since the publication of their report this number has continued to decline. In June 1978 it was down to 77 (Labour Party, 1978, p. 6).

Secondly, declining membership meant that many local parties were dominated by a few individuals. The Labour Party National Executive

Committee carried out a special inquiry into the conduct of the Labour Party in local government in the wake of the revelations about corruption in local authorities involving Labour councillors. The special committee, in its report, was critical of the tendency for local parties and local authority Labour groups to be dominated by a small number of individuals, both party and group often being dominated by the same individuals.

> We are concerned that there is a tendency, particularly in some traditionally strong Labour areas, for party organisation to be inadequate. This leads to the party and the group being dominated by a few individuals and as a result there is inadequate accountability of councillors to the branches.
> (Labour Party, 1975a, para. 11.1)

The special committee did not explicitly make the connection between oligarchy and corruption. Other commentators however, were less restrained. Clarke *et al.* also have some strong words to say on this subject. The three authors speak of 'the rotten boroughs which exist in some areas', and they advocate 'tightening up' which they say would '... reduce the unjust and corrupt practices which can occur where membership is restricted to small numbers and recruitment discouraged' (p. 6). Ron Hayward, the general secretary of the Labour Party, shared their concern. In an interview with *Labour Weekly* he said, 'Some parties don't want new members. They have got a nice comfortable clique and don't want new faces to upset them' (quoted in Hayter, 1977, pp. 14–15). Similar views were expressed by Harold Wilson at the Labour Party annual conference in 1975. He reminded delegates of a statement in his report to the annual conference of 1955. It referred to parties which were 'deliberately kept small for the purpose of concentrating power and local authority seats in existing small cliques'. He continued, 'This has not ended' (Labour Party, 1975b, p. 186). Nor was this problem only twenty years old. A Transport House publication of 1950 says: 'The old-fashioned complacency with a ward committee of a dozen people must give way to a modern conception of a ward association of some hundreds of members' (quoted in McKenzie, 1963, p. 545). This passage, writes McKenzie, 'is clearly directed against a fairly persistent tendency (to be found everywhere in the party) for a small group of party stalwarts to convert the ward organisation into a tightknit band of the faithful who are content to ... manipulate the affairs of the ward without too much interference from a large-scale membership' (McKenzie, 1963, p. 546).

Thirdly, declining membership meant inadequate political activity for party members. Clarke *et al.* are also critical about this.

Many constituencies fail to provide a well constructed programme of activity for the membership as a whole. Too often, general committees become talking shops between semi-institutionalised groups, and members' branches serve no other function than a local government election machine. *The mass party* will never become a reality under present conditions unless membership becomes more than just holding a party membership card. (p. 5, their italics)

PARTY MEMBERSHIP IN NEWCASTLE UPON TYNE

The national picture, then, was depressing. Was the situation in Newcastle similar? The party in Newcastle was organised at three levels. There was a unit of organisation for the city as a whole. Up to January 1976 this was known as the Temporary Co-ordinating Committee (TCC) and after this as the district Labour party. To avoid confusion I shall refer throughout this book to both the TCC and the district Labour party as the district Labour party. The district party was made up of delegates from the constituency Labour parties. After April 1974 there were four full constituency Labour parties and parts of two other constituency Labour parties within the Newcastle boundaries. This was a legacy of local government reorganisation. Below the constituency Labour parties was the third level, the ward parties. They are now officially called branches but in this document I refer to both branches and wards as wards since the latter term was more widely used. There were twenty-six ward parties in total.

Party Membership in 1977
The comments made by some of the party secretaries in giving the 1977 membership totals are more revealing than the bare figures. The Newcastle Central secretary said he was fairly definite about the membership in two of his wards because they had collectors with official collecting books, but he described the other half of his constituency as 'in various stages of decay'. He based his estimate of the membership there on an exercise which the Labour party regional office had carried out in 1976. At that time one of the wards had twenty members, 'all married ladies over 70', he added. The other had forty members. They were, he said, 'people who [the ward councillor] seemed to know . . . from all over the city'.

The secretary for the Newcastle East constituency Labour party estimated that he had 300 members but added that many did not even pay their subscriptions for the full year. 'Someone goes round in July and gets, say, 30p, and that's it', he said. He added, 'I would be surprised if the amount collected is two-thirds of £1.20 (the full subscription)'. He went on to refer to a study which the regional executive committee of the party had organised which had come to the same conclusion.

Academic studies have drawn similar conclusions (Turner, 1978, p. 16), see Table 2.1.

Table 2.1 *Labour Party Membership: Newcastle 1977*

Constituency	Membership
Newcastle North	405
Newcastle Central	290
Newcastle East	300
Newcastle West	603
Castle Ward No. 1	10
Castle Ward No. 2	150
Gosforth No. 1 and No. 2*	82
Total	1,840

*Records kept jointly.

The Transport House publication *Party Organisation* describes 10 per cent of the party's parliamentary vote or 20 per cent of the local election vote as a 'very modest' membership target to achieve (Labour Party, 1972, p. 31). How did the party in Newcastle upon Tyne measure up to this yardstick? As Table 2.2 shows, membership was less than 1 per cent of the electorate and just over 5 per cent of the local election vote.

Table 2.2 *Membership as a Proportion of the Labour Vote and the Electorate in 1977*

Membership	Electorate	Ratio of Membership to Electorate	Labour* Vote	Ratio of Membership to Labour Vote
1,840	224,271	1:122	34,318	1:19

*One ward was uncontested in 1977 and I have therefore added in the Labour vote in this ward for the 1976 election.

Chapter 3

MEMBERSHIP ACTIVITY

How many party members in Newcastle were active members, if 'active' is defined in terms of attendance at party meetings? I have set out below the attendance records of the district Labour party general committee, and one of the constituency Labour parties. This is followed by an estimate of the level of attendance at ward party meetings. Finally I discuss the character of ward party meetings and specifically how they carried out their most important function, selecting electoral candidates.

THE DISTRICT LABOUR PARTY

The district Labour party consisted of a general committee made up of eighty delegates from the constituency Labour parties. It met once every two months. (The Temporary Co-ordinating Committee had met quarterly.) The district Labour party also had a smaller executive committee which met monthly. Attendance of general committee meetings was as follows.

Between March 1973 and December 1975, when meetings were quarterly, twelve meetings were held and the average attendance was twenty-nine. The lowest attendance was twenty. The highest was forty-four (an annual general meeting) that is, just over half the delegates were present. Three of the meetings would have been inquorate under the standing orders introduced later in December 1976. (The quorum was twenty-five.)

Between January 1976 and January 1978 sixteen meetings were held. The average attendance was thirty-one, the lowest was fifteen and the highest forty-eight (an annual general meeting). Between January 1976 and November 1976 three of the meetings would have been inquorate under the standing orders introduced in December 1976. After the introduction of the new standing orders one meeting was inquorate (twenty-four delegates attended) but this fact was ignored, and the meeting took place in the normal way.

Thus the district Labour party was not well supported. It managed without catastrophe to perform its basic duties, namely, maintaining the electoral panel, supervising ward selection procedures and preparing election manifestos; but in doing so it was heavily reliant on a handful of individuals and, in particular, its secretary.

There had been a full-time organiser for the former Newcastle City Labour Party until the late 1960s but when the last incumbent resigned he was not replaced. Since then, voluntary effort had been relied upon. Fortunately, the individual who held the office of secretary for most of the period of this study was an able and conscientious person who in many ways carried the organisation. He resigned in May 1977, partly because of the volume of work involved, and it proved difficult to find a successor. In the end, one of the existing constituency Labour party secretaries, who was also a councillor, took on the job. He held office for a year. Since his departure two other individuals have held the position.

THE CONSTITUENCY LABOUR PARTIES

All the constituency Labour parties in Newcastle relied heavily on the voluntary effort of a few individuals. Not all the constituency Labour parties were organised along exactly similar lines and before examining the attendance record of one it is worth setting out how they were organised. Each of the two part-constituency Labour parties contained two ward parties. Within the Hexham constituency Labour party one ward held regular monthly meetings and one did not meet at all. The two wards within Wallsend constituency Labour party met jointly at monthly intervals. Of the four full constituency Labour parties, three held monthly meetings of their general management committees – as their ruling bodies were called – and one held quarterly meetings. All four had executive committees which met at monthly intervals.

I have set out below the number of delegates who attended the Newcastle North constituency Labour party general management committee in 1975/6, 1976/7 and 1977/8. The number of delegates entitled to attend the general management committee meetings in 1977/8 is shown in Table 3.1.

No records of the delegates entitled to attend were available for the years 1975/6 and 1976/7; the nearest figures available were for 1972/3 when the total number of delegates had been forty-eight. The number of potential delegates was not significantly different in the intervening period.

Attendance was as follows. Between February 1975 and January 1976 nine meetings were held, and the highest attendance was thirty-one (the annual general meeting). The lowest was twenty and the average was twenty-six. Between February 1976 and January 1977 nine meetings were also held. The highest attendance was thirty-eight (the annual general meeting). The lowest was fourteen and the average twenty-four.

Table 3.1 *Delegates to the Newcastle North Constituency Labour Party 1977/8*

Elswick ward	7
Jesmond ward	7
Moorside ward	8
Sandyford ward	6
Wingrove ward	6
Total ward delegates*	34
Women's section (Elswick ward)	2
Young Socialists	2
Trade unions	11
Co-op	1
Socialist Educational Association	1
University Labour Club	1
Polytechnic Labour Club	1
TOTAL	53

*Each ward was entitled to send six delegates plus one more for each additional twenty-five members over an initial fifty.

From February 1977 to January 1978 nine meetings were again held but the attendance is only recorded for eight of them. At the eight meetings the highest attendance was thirty-seven (the annual general meeting), the lowest was twenty and the average was twenty-seven. This figure represents an average attendance record of 51 per cent.

Over the period as a whole about half the delegates were in attendance. The number of members willing to contribute to the work of the organisation was again small. Other studies reveal a similar picture (Cockburn, 1977, pp. 88–9; Forester, 1973). These factors made the constituency Labour party vulnerable to takeover and during the period of this study supporters of the newspaper *Militant*, a Trotskyist sect which dominated the local and national Labour Party Young Socialists, secured a temporary majority of delegates on the general management committee. They did so by taking over one ward party (this was achieved by four or five of them moving house so that they all lived within the ward boundary – indeed for a time they lived in the same house) and by arranging for their own trade union branches to affiliate to the constituency Labour party and to send them as delegates. They gained two additional delegates by securing control of the University and Polytechnic Labour Clubs. They were not the only Trotskyist organisation interested in taking over the constituency Labour party. At one point during 1976 and 1977 there were two such groups vying for power within the constituency Labour party. The International Marxist Group also began to infiltrate the party at this time. However, the latter group made little headway.

During this period the activities of such groups were a subject of concern amongst members of the party leadership. Reg Underhill, the national agent, presented a report to the National Executive Committee on their activities. His report quotes from a document entitled *British Perspectives and Tasks 1974* which did not bear the name of an organisation but was thought to be the work of the *Militant* faction. This document says:

> We must dig roots in the wards and constituencies as we have in the YS [Young Socialists]. Many are still shells dominated by politically dead old men and women. They are now ossified little cliques.[1]

THE WARD PARTIES

From the above it will be obvious that the vast majority of party members did not attend meetings of the constituency Labour parties or of the district Labour party. This left only the ward meetings. As Robert McKenzie pointed out in his study of the Labour Party in the early 1950s, it was only at the level of the ward that individual members had an opportunity for regular participation in the affairs of the party (McKenzie, 1963, p. 547). This was still true of Newcastle in the 1970s.

The Committee of Enquiry into Party Organisation (the Simpson Committee) reported as follows to the Labour Party annual conference in 1968.

> [T]here are not a few parties who succeed in keeping ward committees in active operation in every ward; there are more which have a committee functioning in most of the wards but not all of them; and there are some where the membership is so small that no proper ward organisation can exist at all.
> (Labour Party, 1968, p. 370)

How did Newcastle compare with the national picture described in the Simpson Report? Altogether there were twenty-six wards. They varied a great deal in their level of activity. Some did not meet at all; others met with only a handful of people in attendance; a few held regular meetings attended by more than ten people; one or two held meetings occasionally attended by twenty or more members. It is impossible to be precise about attendance of ward meetings because not all the wards kept records. What follows is therefore an estimate. It is based on my own experience, as a member of three ward parties over the last seven years (as a result of moving house twice) and also as a regular attender of a fourth (the one I represent as councillor at the time of writing). An attendance of as many as twenty was rare, and of as many as fifteen was out of the ordinary. The modal range was six to ten. These figures are representative of other ward parties in Newcastle during this period. From them it is possible to estimate how many party members attended ward meetings in the city as a whole.

Assuming an average attendance of ten members per monthly meeting (a generous figure) this would mean that a total of 260 party members attended meetings of the city's twenty-six ward parties. Allowing for variations in individual rates of attendance over time, I doubt whether more than 400 party members attended ward meetings in any given year: that is, about one fifth of the total Labour Party membership in Newcastle in 1977. If anything, this figure is high. As I have already pointed out, a far smaller number attended the constituency Labour party meetings and a still smaller number the district Labour party meetings.

This result compares favourably with the findings of other studies. Forester found at Brighton Kemptown that 15 per cent of his respondents said they had attended a party meeting in the previous month. Taking into account election activity and meeting attendance he estimates that only between 8 and 9 per cent of the membership could be classified as activists (Forester, 1976, p. 112). Donnison and Plowman in their study of Manchester (Gorton) found that only 11 per cent of members claimed to have attended a ward meeting in the last month, and only 19 per cent in the last six months (Donnison and Plowman, 1954, p. 162). Berry found at Liverpool Walton that 7 per cent of Labour members reported attendance at a ward meeting during the previous month. About a quarter of Labour members helped at election time (Berry, 1970, p. 43).

Not all studies of local Labour parties have drawn depressing conclusions. Turner concluded that the Bermondsey party in 1962 'certainly did not fit the stereotype of a Labour organisation in an impregnable constituency undergoing a process of decay' (1978, p. 318). He found that the Fulham party was a 'healthy well-integrated organisation' (p. 320). However, his findings about the level of activism among members were very similar to other studies. He found in Bermondsey in 1962 that 3.4 per cent of the members were 'active' (p. 159) and in Fulham that 7.2 per cent were 'activists' (p. 160). In South Kensington he found that 8 per cent were 'active members' (p. 160).

Thus only a minority of party members attended ward meetings. And for most of them mere attendance was the limit of their commitment. Most ward parties were kept alive by the hard work of one or possibly two individuals. The overwhelming majority of members were not active even in this minimal sense of attending party meetings. Their commitment to the party was confined to paying their subscription of 10p a month (indeed many were old age pensioners and would have paid only 10p or 20p a year).

Other studies have shown that local ward parties were often only interested to a small extent in issues of policy. They were often more concerned with procedure, with reading the minutes, or with membership, finance and organisational matters, than with discussing political issues (*Fabian Journal*, 1952, pp. 27–32). Batley found in the Byker area of Newcastle that the local Labour party was not 'issue oriented' (1972, p. 103). The situation was similar in many other Newcastle ward parties. Indeed, attending a ward meeting was rarely an experience which encouraged great commitment to and involvement in the work of the organisation. For example, when I first joined the Labour Party in 1972, the ward of which I was initially a member met in a rather dingy room up several flights of stairs. Most of the members were elderly ladies who found the stairs an enormous problem. The chief interest was in the women's section, which met on a separate night and which was mainly concerned with organising social activities such as beetle drives. The meeting room was large and filled with folding tables and creaking chairs. The chairman and secretary sat at a desk near the door and the other six to eight people (sometimes less, very seldom more) sat in a semi-circle around the desk about two yards away from them. In spite of their proximity to the members the chairman and secretary insisted on speaking through a somewhat ancient microphone. At the beginning of the meeting they switched it on and earnestly tested it by blowing into or tapping it. You could hear them if they whispered to each other and yet the microphone was solemnly handed from one to the other as they addressed the meeting. This ward party was not the only one where such details bordered on the farcical.

However, ward parties did vary in character and rather than concentrate on their worst aspects I have given below an account of wards performing their most important function, namely, selecting their electoral candidate. Ward parties were at their best on these occasions.

THE SELECTION OF ELECTORAL CANDIDATES

The responsibility for selecting candidates was shared between the wards and the district Labour party. The ward party chose the candidate but they had to do so from a panel of candidates approved by the district Labour party. How was this panel chosen?

Anyone who wished to be a member of this panel of potential candidates had to secure the nomination of his or her ward party. In all but one case this nomination would normally be supported by the relevant constituency Labour party as a matter of routine. Each person so nominated then had to be interviewed by the district Labour party executive committee. These interviews normally took place during August. Interviews were held in August 1974 (for the 1975 elections), in August 1975 (for the 1976 elections), in August 1977 (for the 1978 elections) and in August 1979 (for the 1980 elections). No sitting councillor was removed from the panel during this time. However, a number of non-councillors were not admitted to the panel. Some of the questioning during these interviews was fairly testing but a major weakness was that candidates were allowed to get away with waffling or ambiguous answers. If questioning became too searching, often someone would come to the rescue of the interviewee. These interviews showed above all that few of the candidates, including many sitting councillors, were well informed about local government. Many failed to demonstrate knowledge of the most basic things. For example some candidates claimed to have a special interest in a particular field. They then failed to show any knowledge of the council's existing policy in that field, or any knowledge of what the local party manifesto said on that subject. The district Labour party executive committee were so appalled by the ignorance of many of the candidates being interviewed in 1974 that they organised a special course for them on the basic features of local government. During the period of this study the non-councillors who were not admitted to the panel were rejected mainly for this reason. However, many sitting councillors displayed similar ignorance and none were rejected. Thus the interviews could be quite testing for non-councillors but were in practice little more than a ritual for sitting councillors.

The interviews were not lengthy. Those held during August 1977 took an average of 12.6 minutes each. This figure is based on twelve interviews held on two separate days. (There was a third evening of interviews which I did not time.) The shortest of the twelve was six minutes and the longest (that involving the leader of the Labour group) was twenty minutes in length. In 1979 the candidates for the panel were interviewed on five evenings. The interviews were timed at fifteen-minute intervals and this was very close to the actual time allowed for each applicant.

Generally speaking, however, the district Labour party executive committee tried to go about its task in a serious and careful way. As studies in other parts of the country have also shown (for example, Cockburn, 1977, pp. 91–2) the same could not be said of many ward parties when they then used this panel of would-be candidates in order to select Labour Party candidates for their wards. What follows is a selection of examples of ward party selection meetings in Newcastle, all but one of which I attended as a party member or as an observer appointed by the district Labour party. The selection of examples is not a random sample. However, the cases chosen do represent a range of ward parties, including both the best and the worst, which I encountered during the period of this study. I have changed the names of the ward parties to conceal their true identity. I have also changed one or two details where these were not important for the essential findings presented below.

(1) I attended the selection meeting of St Cuthbert's ward in November 1974 as an observer appointed by the district Labour party. There were nine voting members present. (This would not have been a quorum under the new rules introduced in December 1976; the new quorum was ten.) The ward secretary did not live within the ward boundary and was not therefore a member of the ward party. She was the daughter of the married couple who both represented the ward on the city council. The ward secretary read out the names on the panel which was then understood to be a requirement. The chairman, who was the third councillor representing the ward, then asked for nominations. The meeting was free to nominate anyone on the panel. Immediately after the chairman had called for nominations an elderly lady sitting on the edge of her seat said, 'Move [name of councillor]'. On hearing this the chairman asked, 'Are there any other nominations?', whereupon the same lady immediately said, 'Move the closure'. On hearing this the chairman said that nominations were closed and that the sitting councillor had been reselected. The councillor then made an acceptance speech.

It was impossible to find out how many members there were in this ward party. The constituency Labour party secretary said that the ward party had never submitted official membership figures to him since 1970 to his definite knowledge. (Before that there had been a different secretary.) There was even some doubt about whether or not official membership cards were issued; certainly the record stubs were never returned to the constituency Labour party secretary.

(2) I attended the St Sigmund's ward selection meeting (for the 1975 elections) as an observer appointed by the district Labour party. The ward party had already held a shortlisting meeting and had shortlisted one candidate, the sitting councillor. The meeting was held in a small hut which had been built to provide a meeting room for the residents of the surrounding estate of purpose-built old age pensioners' bungalows. There were about twenty ladies in the room, all above retirement age. Most came from the bungalows in the immediate vicinity. There were no men and no young women. The chairman, who was somewhat hard of hearing, initially thought that I was present as an opponent of the sitting councillor and it took some time to explain that I was not. Once the meeting started the sitting councillor was reselected within a couple of minutes. This councillor then read out a typed speech, referring at one point to how cosy the ward party was. This ward was also represented by a husband and wife team. They kept this small number of party members happy in order to secure their continued membership of the council.

(3) Gritford ward had been quite lively in 1975 and by 1977 it had built up its membership to more than thirty and had begun holding regular meetings again. But in early 1976, when it had to select its candidate for the 1976 election, morale among the already small number of members was low, mainly as a result of losing the previous election in spite of putting in an enormous effort. Only two members (the chairman and secretary) turned up to the selection meeting. They both lived in the house in which the meeting was being held. There were two candidates to be interviewed. They waited in the back room of the ward secretary's home while the meeting was held in the front room. The two candidates were interviewed, not without a little embarrassment, and one selected. The ward secretary and chairman would have liked the party to be bigger and more active but it simply proved difficult to sustain interest in a ward like this where their opponents kept winning year after year. Nor, however, was Gritford ward the worst in Newcastle. Some wards had difficulty in meeting at all.

(4) Easthope provides an example of a selection meeting in a ward with a somewhat more broadly based membership and where a sitting councillor was being challenged by two other candidates. The Easthope ward meeting was held in a somewhat dingy meeting room above the local Co-op. There was nowhere for the candidates to wait while one of them addressed the meeting. The remaining two had to sit outside in someone's car. According to the ward secretary there were fifty party members. Of these, twenty-seven were present at the meeting. Each candidate was required to address the meeting for up to five minutes and then to submit himself to questions for a further five minutes.

The first candidate told the meeting that he was married with three children and that he was a member of his trade union. He said he would serve the ward to the best of his ability and that he had become interested in people through his union and that this was why he was standing. He finished after about a minute by asking, 'Are there any questions?' and adding, 'Not too hard I hope'.

He was then asked the following questions:

(i)   Are you a Geordie?
(ii) Will you be able to get time off work?
(iii) Will you work hard?
(iv) Will you attend the course for prospective councillors organised by the TCC [Temporary Co-ordinating Committee]? (This was asked by his wife. He answered, 'Yes' – the correct answer.)
(v)   Would you let anyone sway you from your policies? (He answered that he would not, whereupon one of the sitting councillors in the room asked him the next question.)

(vi) Will you follow the group whip? (He answered, 'It depends on what it was. I would be loyal to my people.' This answer was incorrect.)

The second candidate (and sitting councillor) also spoke very briefly, saying that his mother had been a founder member of the local women's section and that he had served as a ward secretary. He concluded by saying which council committees he was a member of.

He was asked:

(i) Would you be prepared to work in wards other than this one?
(ii) Would you attend the course for councillors? (Again asked by the first candidate's wife.)
(iii) Would you be swayed if you thought you were right? (He correctly replied that he would obey the group whip.)

The third candidate mentioned that he had been a member of a neighbouring urban district council for fourteen years and that he was interested in people. He criticised the fact that the area still had no rent office in spite of promises from the council. He said he wanted to see play facilities improved and, in the housing field, that he wanted more old age pensioners' flatlets. He criticised the narrow circulation of the council's official newspaper.

He was asked the following:

(i) Would you attend the course for councillors?
(ii) If there was a by-election in a nearby ward would you help out the candidate?
(iii) Would you hold surgeries?
(iv) Are you a trade union member?
He then left the room and the vote was taken. The sitting councillor was reselected.

(5) Oldene ward had the best-attended selection meeting I encountered during the period of this study. The meeting to select the candidate for the 1978 election was held in November 1977. There were forty-seven voting members present. (Eight more arrived late and were disqualified from voting.) At this meeting the 72-year-old sitting councillor was unseated by twenty-six votes to twenty-one. He was replaced by a young man in his twenties.

I said at the beginning that the above are examples of the best and the worst selection meetings to be found in Newcastle. However, it should not be thought that selection meetings were ranged evenly from one pole to the other. The majority of wards would appear at or towards the worst end of the continuum and Oldene was exceptional. For example, the selection meetings held for the 1978 elections operated under a new set of standing orders and several wards fell foul of them. It was reported to the district Labour party executive committee that three Labour-held seats failed to obtain a quorum of ten members for their meetings (two were highly marginal) and that one Labour-held seat was unable to organise a meeting at all.

The Labour party in Newcastle, then, was large neither in terms of total membership nor in terms of active membership. The party was kept going only by the energy of a few individuals. The selection of electoral candidates – the most important function of the party – was carried out in such a way that the vast majority of Labour councillors were assured of continued reselection more or less regardless of how well they performed their duties. I turn now to the role of the local party in policy-making.

NOTE: CHAPTER 3

1   Quoted in *The Times*, 12 December 1975, p. 5.

———

Chapter 4

THE PARTY AND THE GROUP

Robert Michels showed that even political parties ostensibly deeply committed to democratic principles became oligarchies. For him this was proof of his iron law of oligarchy. Robert McKenzie, in his study of both main political parties, concluded that as long as cabinet government and the parliamentary system were accepted effective decision-making authority would reside with the parliamentary leadership. The leadership had to take account of the views of their organised supporters outside Parliament but in practice final authority rested with the parliamentary party and its leadership (1963, p. 635). Lewis Minkin's more recent study of the Labour Party conference confirms this view. The party was hierarchical although power did not wholly reside with the parliamentary party leadership (Minkin, 1978). Not all students of the Labour Party agree with this view. For example, H. B. Cole, (himself an active Labour Party member) in a more recent examination of Labour Party democracy, disagreed with McKenzie. He concluded that the Labour Party was 'a functioning participatory democracy' (1977, p. 58). However, many other party members seem to share McKenzie's view. Since Labour's defeat in the general election of May 1979 there has been consistent pressure for a shift in the balance of power from the leadership to the rank and file. Three main demands have been made. First, there has been a demand for the party leader to be chosen by an electorate wider than the Parliamentary Labour Party. Secondly, there has been continuing pressure for changes in the party's rules to make it easier to remove sitting Labour MPs. Thirdly, there has been a demand for the election manifesto to be written exclusively by the National Executive Committee instead of by a joint committee representing the NEC and the parliamentary committee of the Parliamentary Labour Party.

There is nothing new about this dispute. Ever since its foundation, the relationship between elected representatives (MPs or councillors) and rank-and-file members has been a controversial issue within the Labour Party. The 1918 constitution resolved some of these difficulties so far as MPs were concerned but the relationship between councillors and local parties was not defined until later. During the 1920s there were a number of disputes between local parties and Labour groups. (See Jones, 1969, pp. 170 – 1.) By 1930 the National Executive Committee found it necessary to formalise the relationship, because many local parties tried to exercise complete control over the behaviour of Labour councillors on local councils. This was felt by the National Executive Committee to be undesirable, although many party members disagreed with them as the debate at the annual conference of 1930 shows. The standing orders proposed at that conference by the National Executive Committee restricted the role of the non-councillor local party representatives at Labour group meetings to a consultative and non-voting one. This proposal was attacked by, for example, the delegate from Bootle, who recommended the practice of his own local party, which met monthly, he said, 'to discuss the council agenda and to instruct the members of the local council as to what they should do'. He felt that the National Executive Committee's proposals did not give local parties 'sufficient control of local Labour groups' (Labour Party, 1930, pp. 163–5; quoted in Gyford, 1976, p. 71). The conference, however, accepted the National Executive Committee's document, which described as 'disastrous' attempts 'to control local administration from outside and to undermine the public responsibilities of councillors'. The document goes on to say that 'under no circumstances must there be any attempt to instruct councillors as to how they should act at group and council meetings'. Nevertheless there were further disputes during the 1930s, for example, at West Ham and Nottingham, as the debate at the annual conference of 1939 shows. However, at that conference the Labour Party approved the standing orders and memorandum (published as 'Labour groups on local authorities') which survived unscathed until November 1975 (when a single sentence was added to the advice contained in the memorandum). Although the standing orders and the official advice removed many areas of doubt a great deal of room for conflicting interpretations remained, and over the years there have been clashes in many localities between the group and the local party.

Many studies of Labour-controlled authorities show that very often councillors have also been party activists (Jones, 1969, p. 173; Baxter, 1972, pp. 103–4; Heclo, 1969, pp. 196–7; Rees and Smith, 1964, pp. 57–8; Butterworth, 1966, p. 28). Rose found that just under half of the individuals who acted as voluntary secretaries within the Labour Party were also elected councillors and that about one quarter of full-time Labour agents were also councillors (Rose, 1976, p. 168). This naturally encouraged internal harmony, but in some cases it also had very damaging consequences. Baxter records that for about twenty years after the Second World War Liverpool was dominated by a very small number of individuals centred around John Braddock (Baxter, 1972). The situation in Newcastle had been comparable to this for the ten-year period up to 1967 when the party was decimated at the polls and lost control of the council. It was during this time that T. Dan Smith dominated Labour politics in Newcastle. Smith was never convicted for corruption in Newcastle but his involvement with the building firm Crudens in the early 1960s aroused many suspicions at the time. Newcastle council began negotiations with Crudens for a housing contract in 1961 but Smith had not declared his interest in the company until July 1962. A price of £772,000 for three high-rise blocks of flats was agreed but vetoed by the then Housing Minister, Sir Keith Joseph. In July 1963 the Newcastle Housing Committee voted by nine votes to one (Smith was no longer chairman) in favour of a public inquiry into the contract. But at the full Labour group meeting before the council meeting due on 31 July 1963 the Housing Committee decision was rejected after a passionate appeal by Smith. Councillor Tom Collins, who at the time of writing was deputy leader of the Labour group, had resigned as vice-chairman of the Housing Committee in a protest against Smith's activities but the Labour group voted against the inquiry at the July council meeting.

The considerable overlap between group members and party activists continued until the local party was reorganised to fight the elections for the new authority due in 1973. Just enough co-operation was achieved to enable them to agree the 1973 manifesto and to fight the elections. But after this the relationship deteriorated badly. The group was certainly dominated by a few individuals; but they did not also dominate the local party. There was some overlap between councillors and party activists but in general few councillors were involved in party activity. This created considerable strains between the group and the party outside the council during the period of this study.

The local party had a potential for influencing the Labour councillors in several ways. First, each councillor had to be appointed to an approved panel of candidates by the local district party and each one had to be selected by a ward party. When their period of office ended councillors had to secure reappointment to this panel and reselection by their ward. Secondly, Labour councillors were expected to attend meetings of their ward party, their constituency party and the district Labour party. At these meetings councillors were made aware of the views of party members. Thirdly, there were formal channels of communication between the group and the party. The district party could appoint delegates to attend the Labour group meetings and the main committee pre-meetings. (Labour members of each committee except Policy and Resources Committee held caucus meetings before the full committee meetings to agree the party line. They were known as pre-meetings. The group allowed the district Labour party to appoint a single delegate to attend these pre-meetings.) Thus, much of the machinery for determining the official Labour line to be pursued in the council or in committee meetings involved the local party, at least formally. Fourthly, the district Labour party had the responsibility for approving the election manifesto which Labour councillors were expected to implement when in power. In this chapter I examine how the formal channels of communication worked and what the formal responsibility of the local party for the manifesto meant in practice. First I consider the constitutional position.

The Transport House document referred to above, entitled 'Labour groups on local authorities', contains a set of model standing orders for Labour groups and a couple of pages of advice about their interpretation. The Newcastle Labour group standing orders were exactly the same as the Transport House model (with one small exception which is not relevant here) and combined with the Transport House advice constitute the formal basis of the group—local party relationship. Two standing orders of the Newcastle Labour group refer to policy-making, and one of the two pages of official advice deals with the relationship between the group and the party. In the standing orders a distinction is drawn between 'election policy' and 'matters coming before the council'.

> The local government election policy of the party shall be determined by the district Labour party in accordance with the terms of its constitution . . . (section 4a)

> It shall be the responsibility of the Labour group on the council to take decisions on matters coming before the council. If the district Labour party desires to express a point of view upon any such matter, it may do so either by communicating with the group secretary or through party representatives at group meetings. (section 4b)

The Transport House guidance on the interpretation of standing orders reads as follows.

> While it is the function of the appropriate local section of the party to determine election policy, it is the definite responsibility of the Labour group to decide group policy and action on the council. Although group policy is determined within the general framework of election policy, questions of practical application continuously arise, and not infrequently decisions have to be taken on matters not specifically covered by election policy. On all such matters it is the definite responsibility of the Labour group to make decisions.
> At the same time, it is proper that the group, in reaching decisions, should consider any views which the local party may desire to express . . . the success of the consultative arrangements depends not so much on standing orders as on a spirit of co-operation and

goodwill among those concerned. While the group must have proper regard to the policy on which the election was contested, the local party must appreciate the responsibility of the group as elected representatives . . . Broadly speaking, since the local party has been responsible for the candidatures of the members of the group, it must entrust to them the carrying-out of Labour policy on the council.
(Labour Party, 1975c)

Thus, formally, ordinary local party members had an important policy-making function to fulfil. The responsibility for determining election policy lay entirely in the hands of the district Labour party. The final responsibility for taking decisions on matters coming before the council was definitely a matter for the Labour group of councillors (though they were expected to consult the district Labour party in doing so). Not only does the party member have an important role constitutionally, the Labour Party also often expresses official pride in its vigorous internal democracy. The foreword to the Labour Party's *Programme for Britain* of 1973 states:

There are two ways of preparing political programmes, of which the one favoured by the Tory Party, whilst not the best is undoubtedly the easier. Not for them the consultation and agreement with those they represent; not for them democracy and debate. Tory policy is developed in private and handed down from the top . . . The Labour Party does not work like this and has no wish to do so . . . Policy in the Labour Party is made by the members.
(Labour Party, 1973, p. 6)

From the above a clear picture of the official attitude, both constitutional and ideological, to group–party relations emerges. I consider below what this relationship was in practice.

ELECTION POLICY

According to standing orders, the district Labour party had the sole responsibility for election policy. It took its role in this respect seriously. For the 1973 elections working groups had been set up; they had prepared proposals and presented them to two policy-adoption conferences. Most of the chairmen of these working groups became chairmen of the appropriate council committees when the Labour group assumed power. Generally speaking, the party election policy then adopted was taken equally seriously by the group and genuine efforts were made to implement most of it.

Those policies had been formulated when Labour was in opposition. Once the group assumed power a gap between the party and the group emerged which grew steadily until May 1977 (when a new group leader was appointed). The district Labour party had set up working groups and held a policy adoption conference in order to agree the election policy for the 1975 elections. But the policies adopted were then almost completely ignored by the group.

It was not until two years later, after the new leader took office, that the 1975 election policies were seriously considered by the group. It must be conceded that it might be that a group will refuse to implement party policies simply because they do not merit implementation. This was not the reason here. This is demonstrated by the fact that when the new leader was appointed and the party's housing document was seriously considered by the group many of the policies were implemented. The group under the old leader had been simply resentful of what some group members regarded as outside interference.

MATTERS COMING BEFORE THE COUNCIL

How did the group and the party deal with matters coming before the council? Here the party was only entitled to consultation. According to Transport House the success of these consultative procedures depended 'not so much on standing orders as on a spirit of co-operation and goodwill among those concerned'. How satisfactory were the group–party arrangements for consultation? In addition to the normal practice of sending letters or resolutions for consideration by the group, there were two formal channels for consultation. These were the party delegates to the full group meetings and (from early 1975) the party delegates to the committee pre-meetings. (From mid-1977 the district Labour party secretary was also allowed to attend meetings of the Labour group executive committee, but this was not a very effective or much-used channel of communication.) Consider first the role of the party delegates to the group meetings.

Party Delegates to the Group Meetings

The district Labour party was entitled to appoint seven non-councillor delegates to the group meetings to put the party's point of view. The district Labour party secretary was an automatic choice and each constituency Labour party or part-constituency Labour party could nominate one delegate. Officially they were supposed to have an important function:

> A ... district party ... must treat the appointment of its representatives to the Labour group as a responsibility of great importance. These representatives must be in a position to attend regularly at the group meetings, to put the party's point of view at such meetings, and to report back to the party.
> (Labour Party, 1975a, para. 11.5)

During the period from October 1973 to May 1977, however, the role of the party delegates did not conform to the pattern recommended by Transport House.

In this respect Newcastle was by no means exceptional nor did it provide the worst example of group–party relations in the country. In some areas party delegates were *formally* debarred from contributing to discussions. The special committee of the National Executive Committee found it necessary to remind some groups that party delegates should be allowed to speak.

> We are concerned that there are some groups in which party represen-tatives are not permitted to participate in discussion. We must make it abundantly clear that attendance in a consultative capacity means that representatives have the right to speak . . . It is outside the spirit and intention of the standing orders for these representatives to be denied this right.
> (Labour Party, 1975a, para. 4.6)

In Newcastle the party delegates were not debarred from speaking. However, in practice they were prevented from making a constructive contribution to the work of the group. From November 1973 to March 1974 I was one of these delegates. This was during the period immedi-ately before the actual takeover of power by the group, when a number of key decisions were being taken in preparation for the handover of power by the old authority to the new.

The group meetings were held on Monday evenings. The agenda and accompanying documents were posted to party delegates on the preceding Friday and normally arrived on the Saturday morning. (Occasionally they did not arrive until Monday or even Tuesday. The lateness and unpredictability were a constant source of annoyance.)

Not all the party delegates were equally interested in contributing to the work of the group. (See Jones, 1969, p. 173.) But four of the delegates began by taking their role seriously. These four party dele-gates met on the Sunday evening immediately before the group meeting with the group chairman (who was also the district Labour party chair-man). He was one of the few councillors who was helpful to the party. The delegates realised that it would be of little use if each of them turned up and merely put their own point of view and so they discussed the (often voluminous) reports in order to try to reach a common district Labour party point of view, which they could then put to the group. To read the reports before Sunday evening and then give up Sunday evening involved a good deal of effort. Initially the four party delegates felt it was necessary.

However, they soon became demoralised about their role. It was impossible for them to make an impact at group meetings. Labour councillors almost always followed their leadership. Indeed, they were reluctant even to discuss the council reports. Most councillors preferred to skip through the whole agenda in two or three minutes simply chanting 'agreed' as the chairman read out the name of each agenda item in turn. ('Chanting' is a fair description.) If anyone interrupted this process, to ask for information or to address himself to the contents of a report, it was usually resented. (The resentment was signified sometimes by words and looks, sometimes by tuts and groans.) If a party delegate (or indeed any councillor) attempted to initiate a serious discussion of a report it would often result in a shout of 'move next business' followed by a prompt chant – again the word is apt – of 'agreed'. Before the unfortunate speaker could pursue his point the meeting would have moved on. Thus although party delegates were allowed to speak, this was only a formality: they were unable to have an effect of any significance on the decisions of the Labour group.

Party Delegates to the Pre-Meetings

I describe below the experience of the party member who was appointed delegate to both the Finance Committee and the Housing Management Committee pre-meetings in the last few months of 1975. His experience was typical of other delegates at this time. An exception was the experience of the representative on the Education Committee, of which the district Labour party chairman was vice-chairman. This prominent party figure tried to be more helpful than other chairmen or vice-chairmen.

Delegates attended the pre-meetings in a consultative, non-voting, capacity. Consultation implies that those involved will be in possession of all relevant information; indeed, it is an essential precondition of genuine consultation. Committee decisions usually took the form of agreeing to a course of action recommended by a chief officer in his report to the committee. If these reports were not made available in sufficient time for them to be read and studied before the pre-meeting, any would-be participant would find himself in a very weak position.

The Finance Committee papers were distributed to councillors on the Friday afternoon for the meeting the following Monday afternoon. But the chairman would not send the party delegate's papers out in the post. Had the chairman posted them there would at least have been the opportunity of reading them over the weekend. Instead, he left them at the civic centre with the leader of the council's secretary for the delegate to pick up – if he could – on the morning of the meeting. If the delegate was unable to do so, then he had to read the reports during the pre-meeting itself.

The party delegate encountered even greater difficulty in obtaining the Housing Management Committee reports. Initially the chairman did nothing, and the delegate had to obtain papers from the Director of

Housing's secretary. She only had a spare copy of the reports of the Director of Housing himself. These normally made up half or less of the committee papers. The remainder were reports of other chief officers; and these the delegate was unable to obtain. Eventually, however, full copies of the reports were made available. The chairman would not post them but again left them to be collected from the leader's secretary. Because of these difficulties party delegates could play very little or no part in the pre-meetings.

The position of party delegates to pre-meetings was very similar to that of the party delegates to the group; they had barely any effect at all on the decisions of Labour councillors. Most Labour councillors looked upon the delegates as being there on sufferance. Apart from the difficulty of getting time off work this probably explains why few of the appointed delegates actually attended the pre-meetings on a regular basis. Indeed in practice several important committee pre-meetings were not attended at all by party delegates.

CO-OPERATION AND GOODWILL

During the period October 1973 to May 1977 there was considerable discontent amongst rank-and-file party members about the state of group—party relations. A campaign to improve them was initiated and in June 1974 the following resolution was adopted by the district Labour party general committee.

This meeting acknowledges the rule laid down in the party's constitution, namely that the party decides election policy and selects candidates to whom it entrusts the task of implementation. On matters coming before council which are not included in election policy the group decides in consultation with the party.

We recognise that there is concern that the reality does not conform to these aims and agree . . . that a working group be set up . . . Its task will be to study the organisation of the council, its committees and party bodies to see if improvements are practicable in all the circumstances, and to report to the next following general committee meeting.

At the meeting of June 1974 hostility was expressed by some councillors to the whole idea of a working group. They did not feel that there was any problem, and argued that the Labour group should be allowed to get on with its work. Two councillors objected more vehemently than the rest to the establishment of the working group and attacked party members for not trusting the councillors. 'There are some people here who do not *trust* the councillors', shouted one prominent Labour group member. Supporters of the proposal replied that it was not a matter of trust. The party could always be improved and should never close its mind to that possibility. There was no need to fear the mere consideration of change which was all that was being proposed. The resolution was carried without its opponents voting against. (They abstained.)

This working group met several times and in March 1975 it put forward some recommendations to improve the situation. The Labour group considered them but made no real concessions. (Slightly before this, however, the group had agreed to the district Labour party proposal referred to above, namely the appointment of delegates to attend committee pre-meetings.)

This is what the district Labour party secretary said about the relationship between the group and the party in his annual report of May 1977:

> *Disagreement between the district party and the Labour group has been a persistent feature of our political life in the last year or so.* On such varied questions as rent review, corporal punishment in schools, election of committee chairmen, powers of the Performance and Review Sub-Committee, sale of council owned houses, use of budget surpluses and reductions in the direct labour force, the party has found itself opposing the group line. In the case of direct labour the general committee warned that it would review its electoral panel with a view to replacing councillors who refuse to support party policy. This decision was an indication of *the frustration felt by committee members at their consistent failure to obtain any real concessions from the group* . . . it is apparent that *a very great deal needs to be done to put internal party relationships onto a satisfactory footing.* This should be a high priority of the incoming officers of the district party and the group. (my italics)

In order to reveal some of the flavour of group—party relations I consider now one of the issues referred to in this document.

The Direct Labour Building Organisation
There was a dispute about the council's direct labour organisation and a row ensued about the district Labour party's right to replace councillors who would not support party policy.

It was strongly rumoured that the group were going to make some of the direct labour building organisation's workforce redundant. There was some support for redundancy in the Labour group and there had been hints by the chairman of the Direct Labour Organisation Committee. As a result a resolution was adopted by the district Labour party at its meeting of October 1976. The minutes record the following.

[T]he secretary reported that at the Labour group meeting Comrade [name of councillor] had strongly advocated sacking employees of the DLBO in order to reduce its size and Comrade [name of councillor] spoke disparagingly about the attitude of its employees to work, describing a policy of no redundancies as 'a charter for laziness'. Following lengthy debate it was RESOLVED:

That this party is opposed to any redundancies in the district's Direct Labour Organisation . . . Should the council initiate any such redundancies, we call upon the district party to re-examine the district panel with a view to replacing councillors who refuse to support party policy.

It was the reference to re-examining the panel that became controversial. In fact a resolution expressing a similar view about the panel had been agreed at the meeting of the district Labour party held in July 1976.

[The] district Labour party is dismayed that in rejecting the party's advice on the election of committee chairmen the district group has once again shown its utter contempt for party policy and resolves on all future occasions when interviewing sitting councillors as candidates for inclusion on the panel to take note of whether they have supported party policy in the past.

This resolution had passed more or less unnoticed. However, the October resolution did not. It was discussed at the following Labour group meeting. The group executive committee discussed the resolution first. The group secretary reported the attitude of the executive committee to the group: 'We are concerned with the clear threat, expressed in the resolution, to take action against councillors for decisions which are taken here [that is, in group meetings]. It was felt to be inappropriate.' He was followed by another member of the executive committee who described the resolution as a 'threat against the democracy of the group'.

Two councillors argued that the resolution was not a threat but merely a reminder of the constitutional position and that the party was at liberty to draw the group's attention to this at any time. To call the resolution a threat was to use inflammatory language and would only make the position worse. However the group was in no mood for reconciliation. One councillor said, 'Well, chairman, this is a threat. We've got the left and the left and the left taking over. We've got to be controlling these people.' Another councillor said, 'OK, we can use words and say this is not a threat. But this is a threat. The district party cannot dictate how councillors should vote nor have they the right to remove councillors from the panel . . . .' (In fact the district Labour party does have the right to remove councillors from the panel, subject to appeal.)

After the debate the group resolved to take the matter up with the Labour Party's regional organiser.

Just how bad general relations between party and group had become is demonstrated by what happened when it was suggested, at a group meeting in 1977, that the surplus income from the group's annual dinner should be donated to the district Labour party's election fund. When the suggestion was formally proposed a heated discussion followed (it was the longest discussion on any item that evening) and the proposal was rejected by fifteen votes to nine.

The above discussion is mainly an account of the group–party relationship up to May 1977. From this time the relationship improved, principally because of the efforts of the new Labour group leader. Until mid-1977 the group–party relationship in Newcastle clearly lacked the 'spirit of co-operation and goodwill' to which Transport House attached importance. The party's effect on policy-making when the group was in power had been negligible; its election policy had been mostly ignored and the consultative arrangements between it and the group meant that the Labour group was able to do more or less as it pleased. After May 1977 the bad feeling gradually disappeared but the district party still fell far short of controlling the Labour group. Its influence on council policy continued to be relatively small; power resided firmly within the civic centre.

Part Two

———————

THE GROUP

Chapter 5

THE GROUP AND DECISION-MAKING

John Gyford suggests three roles which local councillors may play. They may insist that policy-making is a matter for the elected member and devote considerable time and effort to it: policy initiation. Or the councillor may leave the initiation of policy largely to officials and concentrate instead on examining the proposals brought forward by officers: policy scrutiny. Or, thirdly, he may accept uncritically the policies advocated by officials while acting as the guarantor of the proposals in the party group or in council or committee meetings. Such individuals, suggests Gyford, derive their satisfaction 'not from the content of the policy so much as from being "in the know", "pushing things through" and "getting things done"'. He dubs this role 'policy acceptance' (Gyford, 1976, pp. 136–7).

These categories – policy initiation, policy scrutiny and policy acceptance – may be usefully applied to analysis of the way in which Labour groups as a whole go about policy-making. In describing the way in which the ruling group at Newcastle upon Tyne functioned, I have asked three questions. First, to what extent did the Labour group either initiate policy or act as a setting in which individual councillors could initiate new policies? Secondly, how effectively did the group act as a setting for the scrutiny of official proposals? Thirdly, to what extent could it be said that the group's role was one of acceptance?

POLICY INITIATION

Bulpitt studied six local authorities in the north west of England and found that party groups did not function as effective policy-making bodies. He found 'few signs to indicate that any such group was particularly successful in its efforts' (1967, p. 121). Jones's study of Wolverhampton concluded that the majority Labour group was a 'hand to mouth affair living from agenda to agenda' (1969, pp. 175–6). Gyford concluded that it was 'all too easy for groups to content themselves with a quick run through the next agenda and to avoid the complexities of policy planning' (1976, p. 79). Other Labour groups, however, had a little more regard for policy development. In Leeds, for example, there was an advisory committee which acted as a source of ideas (Hill, 1967, p. 146). What was the position in Newcastle upon Tyne?

Transport House urges local Labour groups to discuss forward policy questions. The Labour Party's *Local Government Handbook* says:

> It is important to ensure that all groups regularly discuss long term policy. The practice of holding two meetings in each committee cycle, whether one meeting is reserved specifically for discussion of long term policy questions, or whether there is the discussion of such questions at both meetings, will facilitate this.

The document says that in some authorities Labour groups held full group meetings before the monthly meetings of their Policy and Resources Committee and that other groups regularly allocated another date in the council cycle for consideration of general policy matters (Labour Party, 1977b, p. 348). Newcastle upon Tyne compared badly with these other authorities. During the municipal years 1976/7 and 1977/8 only two special meetings were held to discuss future policy questions.

Some Labour groups elect a policy advisory committee or executive committee charged with the task of developing forward policy and bringing proposals to the full group for approval. Indeed this was the system recommended by Transport House in its model standing orders (Labour Party, 1975c). The Leeds Labour group had its own policy advisory committee which met on alternate Mondays as well as before the mid-monthly Labour group meetings (Wiseman, 1963a, p. 67). The Labour group in Newcastle upon Tyne had an executive committee which was supposed to consider issues of policy and to advise the group accordingly but its role in this respect was very limited. It normally only met for one hour immediately before the regular monthly group meetings. Special meetings of the Newcastle upon Tyne Labour group executive committee to discuss forward policy questions were rarely held. This made it virtually impossible for it to achieve anything of significance.

Before local government reorganisation the executive committee had played a more important role in policy development. But during the transition year between the old and new authorities (1973/4) – when such organisational issues were settled – the group had failed to consider adequately what roles either the group or its executive committee were to play in policy-making and policy development. As a result the group's functions had been taken over by the council's official Policy and Resources Committee. As one backbench councillor said during interview: 'Policy and Resources Committee tend to treat themselves as an executive committee of the group. They displace the executive committee in initiating and presenting policy . . . the group loses out.'

The group and its executive committee did not, then, play a significant part in initiating policy decisions. Nor did they provide a setting in which individual councillors could put forward new proposals for serious consideration. How effectively did the Policy and Resources Committee function?

The Policy and Resources Committee, unlike all the other committees, was a one-party committee. Its chairman was the leader of the Labour group (and the council) and its vice-chairman the deputy leader of the group (and the council). Its terms of reference gave it wide powers to lead and co-ordinate the work of the whole council: it controlled the council's capital programme, and was charged with reviewing the effectiveness of all the council's existing services and with identifying the need for new ones. Yet, in spite of its tremendous importance, no one I spoke to felt that the Policy and Resources Committee worked satisfactorily.

Labour councillors drew attention to what they regarded as seven main weaknesses. First, that on only one occasion (while considering the 1974/5 budget) did it meet without all the chief officers present. As one main committee chairman and Policy and Resources Committee member said:

Policy and Resources Committee does not work well at all. There should be a group meeting to discuss our feelings openly before the meeting. As it is we have to argue amongst ourselves in front of the officers. This inhibits discussion . . . At present it's just a receiving body. It should initiate more.

A second criticism was that the Policy and Resources Committee was too large for effective discussion. During the period of this study it had between fourteen and eighteen members. In addition, a dozen or more officers attended. This meant that a minimum of twenty-five to thirty people were present at its meetings.

A third criticism was that Policy and Resources Committee was dominated by officials. This problem was more serious under the old leader. Councillors felt that he exercised inadequate control of the agenda and reports. Until May 1977 control of the agenda and the reports was at best slender and at worst sporadic and unpredictable. The new leader however did carry out these functions more effectively. In short, when the new leader chose to be in control he was able to do so. However, he tended to limit severely the number of issues which he got involved in. This meant that, in practice, officials were still able to secure the support of Policy and Resources Committee very easily.

This is how one main committee chairman referred to this problem:[1]

> The policies of the council are emanating from a few officers. These are then agreed by Policy and Resources Committee and then by council. It was agreed that Management Team[2] would present alternatives to committees and not just a single recommendation. I cannot recall this happening once ... The officers make the agendas too long. The members all sit with their eye on the clock. Important decisions can be made without discussion. There may be discussion early in the meeting but not as it wears on. When we consider reports the chances of reversal are very small ....

This councillor was also interviewed in autumn 1977. He felt that there had been some improvement as a result of the change in the leadership in May 1977 but not as much as he felt was necessary.

The Management Team, during the period immediately after local government reorganisation, functioned very much like a council committee, receiving reports and taking votes on them. Its members also adopted the practice of taking collective responsibility for its decisions contrary to the wishes of the Policy and Resources Committee, which had agreed that when officials presented policy documents to Policy and Resources Committee they should set out alternatives. Furthermore, Policy and Resources Committee members were unable to obtain all the reports considered by the Management Team. They were also unable to attend its meetings:

> When it was set up I argued that we should attend. [The then Chief Executive] said OK at first but gradually it became 'No'. I didn't press the issue to a vote in Policy and Resources Committee because I knew we would lose. Not enough members would support me ... When the issue of councillors attending was raised the Chief Executive said that if we did go [to the meetings] they would just meet somewhere else – in the toilet if necessary. He added that he would not mind me going but that there were some [councillors] that they didn't want.
> (a Policy and Resources Committee member)[3]

A second main committee chairman also criticised the influence of officers on the Policy and Resources Committee: 'The officers are pleased with Policy and Resources Committee because it's in their pocket. The officers often present reports with contempt.' When asked how the officers were able to get away with this he replied, 'Mainly because the politicians are weak. We need to get the right balance of individuals on Policy and Resources Committee. They are often frightened. If we had the best people in the group there, you would get a different atmosphere.'[4] He supported his assertion about officer domination by referring to the way in which the corporate plan working group had been set up. This was a working group made up of the leader and a small number of leading chairmen. It was a further centralising factor and strengthened both the power of the leader and of a small number of senior chief officers. Its main impact, however, was confined to the preparation of the annual budgets. It was set up on the initiative of the Chief Executive in 1974, partly because of his need for clear and consistent political guidance which the then leader failed to give. The chairman said that the officers had had a big hand in deciding who would be put on the working group, and continued:

> The officers are now dealing patronage out. They decided who should be on corporate plan. They had a word with me, the leader and with the group secretary and then presented the list. There was a joint report of the leader and the Chief Executive, giving the names. They play the same game as us. They know politics and what makes it tick.

A fourth criticism was that personalities played too big a part in Policy and Resources Committee's proceedings:

> The chairman [the leader] is well prepared but he's dealing with some very awkward characters. Decisions are not always taken on logical grounds but rather because of the force of personalities.
> (a main committee chairman)

A fifth criticism was that many of the committee chairmen on Policy and Resources Committee acted purely in the interest of their own committee:

> There are easier processes for getting something done than having a go at Policy and Resources Committee . . . It's difficult because there are so many mainline committee chairmen looking after their department. They are reluctant to oppose anything to avoid creating hostility to their own proposals.
> (a main committee chairman)[5]

A second main committee chairman admitted that he did this himself and gave the following reason: 'Policy and Resources Committee is dead boring. I go because I have to as chairman. I find [my committee's] work interesting and stimulating and concentrate on that.'

A sixth criticism was that few members of Policy and Resources Committee knew enough about most of the issues to contribute to the discussion. Some Policy and Resources Committee members could not, or would not, read all the reports.

The timing of the distribution of papers did not help this problem. The regular meeting time was on Wednesday afternoons (at monthly intervals). Most of the papers were distributed the Friday before but large numbers were circulated only the day before the meeting and some were put round the table. This gave members four and a half days in which to read the papers distributed on the Friday and half a day to read papers circulated on the Tuesday. It allowed two and a half days to speak to officers about reports distributed on the Friday and only half a day to discuss those circulated on the Tuesday.

One subcommittee chairman, interviewed in 1977, said he had difficulty in getting through all the committee reports. Another subcommittee chairman said he also had the same problem: 'The reports are so voluminous you're unable to deal with them properly. You need somewhere quiet to read through. It's impossible to cope, to get through this enormous volume of papers.'

Policy and Resources Committee members that did read the documents relied heavily on these officers' reports for their knowledge of the facts. This is demonstrated by the Policy and Resources Committee members' answers to questions 29 and 30. Nine out of fifteen of them said they gathered either no extra information before meetings or that they spent only a little time gathering extra information before meetings.

A final criticism was that many of the Policy and Resources Committee members did not put in sufficient effort. One leading member said that most members had very little effect on its decisions, adding: 'Some don't speak. They don't use the power they've got. They don't *try* to use the power they've got. How much effect you have depends on the use of initiative and skill.'

The consequence of these difficulties was that reports – often on issues of major significance – tended to be agreed without anything approaching a thorough discussion or examination by councillors. As one subcommittee chairman put it: 'I was on Policy and Resources Committee for a long time. To me it was a dead-end thing. Everything presented to it used to go through.'

Newcastle's was not the only Policy and Resources Committee to work ineffectively during this period. Professor J. D. Stewart has been very critical about how most of the Policy and Resources Committees, set up in the wake of the Bains Report, have functioned.

> Policy and Resources Committees are drowning in a sea of paper. One meeting a year, but lasting for a week and looking at policy in depth, would be better than the endless grind of committees to which you have subjected yourself (*sic*) and the rigid agendas in

which you have allowed yourselves to be imprisoned.
(quoted in Young, 1977, p. 126, n. 74)

The failure of the Policy and Resources Committee to give adequate consideration to the issues which were brought before it, or to evolve a satisfactory way of considering questions of future policy, was particularly serious in Newcastle. As I pointed out above, before the reorganisation of local government the task of developing future policy had been carried out by the Labour group under the guidance of its executive committee. Its role had been assumed by the council's Policy and Resources Committee. The ineffectiveness of this committee meant that outside the mainline committees systematic thinking either about key issues of the present or about future problems was not taking place among majority party councillors. I consider later how effectively the main committees functioned.

POLICY SCRUTINY

How effectively did the Labour group of councillors deal with policy proposals which had been initiated elsewhere? In this section I examine how the group considered the monthly council agenda, how it dealt with the annual budgets in 1976/7 and 1978/9, and how it dealt with the housing investment programme bid for 1977/8.

The Council Agenda
Hill found that there was 'quite a lot of "referring back"' at the meetings of the Leeds Labour group held immediately prior to the council meeting. Chairmen answered questions from members 'on a formal basis, with written questions', although supplementary questions were not generally allowed (Hill, 1967, p. 147). Wiseman felt that the Labour group in Leeds overruled the committees too frequently (1963a, p. 65). He reports that examples were 'too numerous to cite' (Wiseman, 1967, p. 95). How did Newcastle upon Tyne compare?

At the regular monthly group meetings, called principally to discuss the business of the next council meeting, there was a standard agenda:

(1) Apologies for absence
(2) Minutes of previous meeting
(3) Matters arising
(4) Group leader's report
(5) Chairman's question time
(6) Council agenda
(7) Correspondence
(8) Any other business

From time to time special items were also added to the list. Chairman's question time had been included because a restrictive attitude was adopted towards Labour councillors asking questions at the council meetings: they were expected not to ask questions which might embarrass the party in any way. Because this put Labour councillors in a weaker position than opposition councillors, Labour councillors were allowed to ask such questions at the group meeting. There were almost always some questions asked by group members but the question asked at a group meeting was a far less effective instrument than the question asked at the council.

Before discussing how the group dealt with the council agenda it is worth referring to a regular limitation on its effectiveness. Normally the group meetings were held on the Monday preceding the council meetings at 7.15 p.m. The small number of special meetings were also normally held at this time. The papers for these meetings were normally distributed on the Friday before. This meant that most Labour councillors received them on Friday evening after work. This left only three days to read any documents. More important, it only left a single day on which the civic centre was open, and on which officers could be contacted in order to check information in the papers or to gather extra information which councillors felt they required. These problems put Labour councillors at a considerable disadvantage.

The full council meeting is the sovereign body of a local authority. The main business of every Newcastle council meeting was to consider the reports of the committees. The vast majority of the committee reports had the status of recommendations which required the confirmation of the full council. (A few were for debate only. The committees were not required to seek the approval of the full council for all their decisions; they all had considerable powers directly delegated to them, but they sometimes submitted decisions for which they had delegated powers to the council 'for information and discussion'.) In Newcastle upon Tyne – as in many other local authorities – the meetings of the full council were largely ritualistic in character. This was because the majority and opposition party groups voted in council according to their own predetermined line. How did the controlling Labour group go about agreeing its official line?

Formally the group was entitled to reject any report on the council agenda but in practice it rarely did so. It considered the items in a highly perfunctory manner. I measured the length of time spent on the council agenda by seven group meetings in 1978. The results are set out in Table 5.1.

On average, forty-four seconds were devoted to each committee report. In fact this figure overstates the average time devoted to discussion of the reports, as column (4) in the table shows. At four of the meetings the greater part of the time devoted to the committee reports

Table 5.1 *Time Spent by the Labour Group Discussing Committee Reports to Council*

Meeting/ Average	(1) Number of committee reports on agenda	(2) Time spent on the committee reports	(3) Average time spent on each committee report	(4) Time spent on a single committee report (% of total time spent on commit- tee reports)
Meeting 1	13	25 minutes	1 minute 55 seconds	15 minutes (60%)
Meeting 2	20	15 minutes	45 seconds	13 minutes (87%)
Meeting 3	16	7 minutes	26 seconds	— (—)
Meeting 4	19	12 minutes	38 seconds	7 minutes (58%)
Meeting 5	13	1 minute	5 seconds	— (—)
Meeting 6	18	24 minutes	1 minute 20 seconds	21 minutes (87%)
Meeting 7	31	11 minutes	21 seconds	— (—)
AVERAGE	—	14 minutes	44 seconds	—

as a whole was devoted to a single report. (At these meetings one or more councillors attempted to debate a particular report. These attempts came to nothing.) If the time devoted to these four reports is subtracted from the total then the average time spent on the remaining 126 reports works out at nineteen seconds. Furthermore the nineteen seconds overstates the actual time spent on the vast majority of the reports. Most group meetings were characterised by a marked reluctance to discuss the committee reports at all. Business was rushed through; attempts by councillors to raise issues were largely met with silent resentment. The attitude of many group members was typified by one councillor who objected to the fact that the group chairman kept reading out the names of the reports which were to be submitted to council. He felt that it took too long to read out the name of each item and suggested that, instead, the chairman only read out the number.

One councillor described what usually happened if a councillor was dissatisfied with a committee decision and tried to raise it at a group meeting.

In the main the chairmen . . . want to keep you subdued . . . You could resurrect something at group. It might get a proper airing there. But the way the group's balanced . . . you would not get a fair hearing there either. They think you're a troublemaker, just stirring. The chairmen always support each other, and the vice-chairmen. The backbenchers don't see themselves as backbenchers. How many are able to sort information out which they get from the officers? Only about a third of the group are probably aware of what they can do . . . I'm known as a left-winger and so I can't make any progress. Even after weeks of preparation by a backbencher he still wouldn't prevail.

During the group meetings I have attended as a councillor no report on the council agenda was referred back or rejected. (Very occasionally a report was withdrawn on the request of the appropriate committee chairman.)

Before the reorganisation of local government the group in Newcastle had dealt with the council agenda more effectively. One group member, a subcommittee chairman who had been a councillor for over twenty years, compared the group before and after reorganisation in the following terms.

[Before reorganisation] items were brought to the group more often. They were properly discussed. And changed. They were more thoroughly discussed. I was off the council . . . I came back on in 1973 and noticed the change then. Group business was more hurried. Previously we got the council agenda and properly discussed the committee reports in group meetings. Unlike now. It just goes through . . . You've got to do a lot of lobbying, but things are often dismissed anyway.

A second councillor who had served on two other authorities said that in Newcastle in the past 'the group often amended a chairman's report'. The group, he said, received reports 'not at the stage of the agenda like now', instead, reports 'were considered before the agenda was made up'.

The obvious explanation for the peremptory way in which the group dealt with the council agenda was that there was no need for it to go into the reports in detail because the committees had already done so. This reasoning is not without validity. But, the weakness of this argument is that the committees did not go into issues in sufficient detail either. However, I examine the committee decision-making process in a later chapter.

A second explanation for the limited time devoted to discussing reports could be that the reports simply did not warrant it. This reasoning is also not without validity. Many of the reports were recommending approval of particular contracts which were already included in the local authority's programmes. To an extent they had already been 'approved'. However, it was not merely reports of that kind that were given such rapid consideration. Reports were handled in the same way more or less regardless of the importance of their contents. Even a decision as important as the housing investment programme submission to the Department of the Environment was not immune from such handling. Nor was the all-important annual budget decision.

The Housing Investment Programme
The council was required to submit an annual bid for housing funds to the Department of the Environment. Once its allocation was fixed it would not normally be allowed to spend (apart from the tolerance arrangements which anyway depended on the total sum allocated) in excess of the figure allocated. Most aspects of housing policy were affected: new building, slum clearance, modernisation, mortgage lending, lending to housing associations, improvement grants, and so on.

The decision for the financial year 1977/8 was taken and the bid of several million pounds submitted to the Department of the Environment without the knowledge of the group or the appropriate committee. The submission was presented to the full Labour group (and to the Housing Committee) after the document had been sent off to the Department of the Environment. Worse still, the document presented to the group was badly printed and it was not possible to read all the figures. A small number of councillors referred to this problem at the meeting. For example, a highly experienced backbencher said, 'You can't expect anyone to take an intelligent interest in something printed like this – that you can't read'. Most group members, however, said nothing.

The decision had been taken in isolation from other councillors by the Housing Committee chairman along with the Housing Management Committee chairman and the leader (in their capacity as members of a new subcommittee of the Policy and Resources Committee called the housing policy group). It might be said that there was no need for it to be presented to the group or to the committee because the information was straightforward and that therefore the task of submitting the bids could be safely left to council officers and to the new subcommittee. This was not so.

For example, the document contained a statement of the council's overall housing strategy. Clearly such a statement should have been discussed and agreed by the appropriate committee and by the Labour group. However, the strategy statement was prepared by officers and approved only by the two chairmen and the leader.

Secondly, the document contained bids based on assumptions which were not discussed by the group or the committee. An assistant director in the Housing Department described one figure as 'plucked out of the air'. In this case an initial figure had been suggested by the Housing Department and increased in the City Treasurer's Department 'by a purely arbitrary amount', he said.

Thus, this all-important decision was given even less consideration than the council agenda. There was not even a token opportunity to reject or amend the document.

The Annual Budget

I turn now to an examination of how the group dealt with the key issue of each year. The budget fixes the amount of money to be made available to committees to pursue their objectives. It fixes the framework within which some priorities must be determined and in the vast majority of instances it actually determines what these priorities will be. It is the most political of all local authority decisions. Therefore investigation of the way a Labour group deals with its annual budget constitutes a good test of its performance and effectiveness. I describe how the group went about agreeing two annual budgets, those for the financial years 1976/7 and 1978/9. (I have left out the decision concerning the 1977/8 budget. Because that budget was dominated by central government controls the group was not free to move substantially one way or the other. Examination of how the group reached its decision in that year was therefore not particularly revealing; it threw light on the extent of central control rather than on how the group arrived at decisions.)

The 1976/7 budget. Papers for the meeting were not sent out on the Friday before as they normally were, but were distributed just before the meeting. Group members received a note from the group secretary on the Friday saying: 'Please note that papers appertaining to the budget will be distributed half-an-hour before the meeting on Monday night, in order that you may peruse them.' The reason for this, the leadership claimed, was to prevent leaks to the press. It had always been suspected that someone in the group leaked information to the press and much was made of this in order to justify secrecy. In fact the contents of the document appeared in the local evening paper on the Friday before the group meeting. This was how most Labour councillors learned for the first time that, against all expectations, a reduction in the rate was being proposed. Furthermore, the papers for the group meeting was not available even for the full half hour because the chairman of the Finance Committee arrived with the reports about five minutes late. The document was printed on eight sides of A4 paper and it was impossible even to read through it, much less to understand it, in the few minutes available.[6] The short notice meant that there was no chance to think about the proposal over the weekend, no time to weigh the evidence in the document (such as it was) and no time to gather any extra information which councillors found necessary on reading the report. Some Labour councillors resented this but most did not. Indeed only a handful of group members turned up half an hour before the meeting to peruse the document.

The main proposal was to cut the rate by two and a half pence in the pound. The council had balances available which were several million pounds more than had been anticipated, and this provided the opportunity to reduce the rate. The debate at the meeting was poor but an amendment to the proposal was moved: 'That the domestic rate be held at its present level and that the balances be spent on housing maintenance, the "priority areas" and other socially worthwhile projects.' The amendment was put to the vote, but only four councillors voted for it. Whatever the merits of the decision, it was taken by councillors who had not had the opportunity to read the document and without proper discussion.

No detailed work on the budget had been carried out by the group. Nor had such work been carried out by its executive committee. It was done by a subcommittee of Policy and Resources Committee, the corporate plan working group. Circulation of its documents was highly restricted and its meetings held in secret. (In Leeds, by contrast, the policy advisory committee of the group – the equivalent of the executive committee in Newcastle – played a very much larger part, including detailed scrutiny of committee estimates (Wiseman, 1963b, p. 147)).

The 1978/9 budget. The 1976/7 budget decision had been taken at a time when the group was dominated by its leadership. The 1978/9 decision was taken in different circumstances: there was a new leader and there was organised opposition to him.

An initial group meeting was held at which the basic strategy was agreed. This was essentially to allow a small amount of growth in services, concentrated in fields which had been neglected in the previous four budgets. (No strategy meeting had been held during the period of office of the former leader.) The details of the budget were then worked out within the framework of this strategy and brought to a second group meeting. There was a long debate at this second group meeting. It was also one of the best debates I witnessed. There was opposition to the 13.5 pence rate increase which the group executive committee were recommending. A vote was taken and the executive committee proposal was agreed. After the issue had been settled, however, a move was made by supporters of the former leader to reduce the extent of this increase. About three weeks later these opponents won a vote (against the will of the leader) at a group executive committee meeting, by six votes to five, to reduce the increase from 13.5 pence to 11.5 pence. This reduction was then recommended to the full group which, at a third meeting, voted in support of the executive committee recommendation (to reduce the extent of the rate increase).

This example shows how important control of the executive committee could be. The leader's power rested to a considerable extent on his ability to secure majority support in the executive committee. The former leader had experienced little difficulty in doing so but the new leader was not so fortunate.

POLICY ACCEPTANCE

It is time to summarise the position so far. The group did not play an important part in the initiation or development of policy. Nor did it effectively evaluate proposals which had been initiated elsewhere. Its role was much closer to the acceptance role in Gyford's usage.

The Labour group was the sovereign body for Labour councillors in Newcastle. When it took clear-cut decisions they were not openly defied. But this did not mean that the group itself was an important forum for actually taking decisions. The group acted principally as a receiving shop, serving to legitimise decisions which, in reality, had been taken elsewhere. This was confirmed by the answers Labour councillors gave during interview. They were shown a list of institutions and asked to say, first, which had the 'most power' (question 70) and, second, which had the 'second most power' (question 71).

Thirty-nine Labour councillors answered the first question. Ten (26 per cent) thought the leader of the council had the most power. Thirteen (33 per cent) thought the Labour group did. Only two thought the Management Group of officers had the most power but eight (21 per cent) thought the Executive Group of officers did. A further six councillors (15 per cent) thought that the Policy and Resources Committee had the most power. Thus, 67 per cent of Labour councillors thought that other bodies had more power than the Labour group.

In their answers to the second question, only five councillors who failed to mention the group as having the most power ranked it second. Twelve Labour councillors (32 per cent) thought the leader of the council was second in power and a further twelve (32 per cent) thought that either the Executive Group of officers or the Management Group of officers were second in power.

If the answers to both questions are taken together then a total of seventy-six responses were given. The Labour group was mentioned as having the most or the second most power on eighteen occasions (24 per cent). The leader of the council was mentioned most often, on twenty-two occasions (29 per cent). The Executive Group of officers plus the Management Group of officers were also mentioned on twenty-two occasions. Thus, 76 per cent of the responses rated other bodies more highly than the Labour group. The majority of Labour councillors did not think that the group had either the most power or the second most power.

Disunity

Two underlying factors explain the weakness of the group in the council power hierarchy. First, the Labour group was not united. Secondly, there was a tendency for the group to be dominated by a small number of individuals centred on the leader. Real power resided where they allowed it to reside. The issue of oligarchy is the subject of Chapter 6. The issue of disunity is discussed below.

Gyford notes a tendency for conflict within Labour groups (Gyford, 1976, p. 84. See also Janosik, 1968, pp. 85–112; Bulpitt, 1967, p. 42; Bealey *et al.*, 1965, pp. 405–6). A major criticism advanced during interview by Newcastle's Labour councillors was that the Labour group was not united, that there was too much disharmony. The differences were both personal and political in nature.

A prominent executive committee member said that one of his greatest sources of dissatisfaction was 'the squabbles with our own members'. 'Sometimes', he added, 'the jealousies and viciousness of what goes on behind the scenes, sometimes this sickens me.' A main committee chairman – speaking of the period prior to May 1977 – said that he found personal conflicts within the party a source of dissatisfaction. 'For example,' he said, '[a named senior councillor] threatened to throw me out of the window for putting [a named committee's] estimates up when he wanted to cut them. You get better treated by the officers.' Amongst officers, he added, there was a tradition of courtesy towards chairmen. A main committee vice-chairman said that he found 'a lack of comradely support in the group; a lack of fraternal attitudes'. He continued, 'You get little help. Some are thrown into petty jealousies. There are people who prefer to keep good people out.' A second main committee vice-chairman said, 'We don't always pull together as socialists. People fall out for personal reasons rather than over the programme.'

Other councillors stressed the political differences. One backbench councillor said, 'There are many in the party whose spiritual home is the Tory Party. Many of them don't want to change the system.' An elderly backbench councillor talked of 'the disharmony in the group'. 'Many times I think I'm in a Tory place', he said, 'we should be together more.' Other comments were:

I'm disappointed in so many things . . . I expected change in 1973 but any progress has been so slow. You don't always have the help of your Labour colleagues, quite honestly. I thought Labour was

socialist and the Tories, Tory. I'm disappointed with a lot of colleagues. It's such a hard slog to make progress. I thought I would get more support. I thought we were all going along the same road. But we're not. I've been very disappointed and disillusioned at times.
(a main committee vice-chairman)

We would have made more progress if we were like-minded. So many colleagues hinder you. They're at cross purposes.
(a subcommittee vice-chairman)

The Group and Officers

The Labour Party's *Local Government Handbook* reports that a number of groups arrange for council officials to be present at the beginning of meetings to give factual information and advice. The officers are requested to leave before these meetings begin to reach political conclusions (Labour Party, 1977b, p. 349). The Newcastle Labour group did not normally meet with council officials present. From January 1976 to October 1978 officials were invited to only three meetings. One official was invited to a meeting to explain the Community Land Act. Questions were put to him and he then left before the debate began. On the second occasion, an official was invited to a special meeting called to discuss a local green paper on priority areas. He introduced the report and remained during the debate but he took no part in it. On the third occasion, two officers were invited to discuss a report on personnel management. They also remained in the meeting during the debate but took no part in it. The officers' role on each occasion was confined to providing information and answering questions.

NOTES: CHAPTER 5

1   This interview took place in November 1974 and was carried out by the district Labour party's organisation working group. All subsequent references to interviews with councillors carried out in 1974 concern interviews carried out by this working group.
2   The Management Team of officers. This was made up of all the chief officers and was led by the Chief Executive. It was later called the Management Group of officers.
3   Interviewed in November 1974.
4   Interviewed in November 1974.
5   Interviewed in November 1974.
6   The full budget document, listing each committee's proposals in detail, was not presented to the group. Instead, members received a copy of the proposed report to the council which contained a summary of the budget, offered some argument in favour of supporting it and ended by recommending approval of the full budget document.

Chapter 6

THE POWER OF THE GROUP LEADER

For Michels an all-powerful leadership was inevitable. For the Labour Party it was a danger to be avoided. This is the advice given to members in the party's official *Local Government Handbook*.

> Any tendency for power to be concentrated in a few hands should be resisted by all members of the group and by all members of the party. Although in any given situation particular individuals might be acting perfectly correctly, they might, because of their personal qualities or seniority, come to dominate both the party and the group. This will lead to a narrowing of political vision, unimaginative policy making, and ill-considered decisions. The resulting disillusion with the process of Labour Party policy making would deter party members from seeking elected office, and deter members of the public sympathetic to our basic objectives and interested in affecting local council policy from joining the party.
> (Labour Party, 1977b, p. 352)

How successfully did the Newcastle Labour group avoid these dangers?

In the days of Mayor Richard Daley the Chicago City Council became known as 'the nod squad' because of their blind subservience to him. During the period up to May 1977 the Newcastle Labour group equally deserved this appellation for their uncritical support of their own leadership. This description would have applied to many local authorities in this country. This fact enabled T. Dan Smith and Poulson to operate as they did. Smith thought that the vast majority of councillors were not worth bothering with. He said, 'If 95 per cent of them never went near the council chamber it would make no difference'.[1] He preferred, he said, to go for the 'strong men'.[2]

Some Labour councillors referred to this problem during interview. For example:

> I'm not happy with decision-making among a small inner group. This was initiated by [T. Dan] Smith. Someone . . . wrote a book . . . He said that you were either on the stagecoach or you were not on and

were trying to get on. Smith organised things that way. There were hangers-on and people who he could give a bit of a nod to if he wanted support in a meeting.
(a main committee vice-chairman)

'My main source of dissatisfaction', said a subcommittee chairman, 'is the operation of cliques which dominate the power structure in the council.'

At the centre of the cliques referred to above was the group leader. The following factors made it possible for the Newcastle Labour group leader to exercise a dominant influence on the affairs of the Labour group.

The most important factor was the power of patronage possessed by the leader. His position was reinforced by the way in which he was able to utilise the convention of collective responsibility which applied to members of the group executive committee and the council's Policy and Resources Committee. These factors are examined below. The leader could also use the party's disciplinary machinery to his advantage, but this is the subject of the next chapter.

COLLECTIVE RESPONSIBILITY

The fifteen-strong executive committee was elected at the group's annual general meeting. Elections were held for the following positions at the annual general meeting in May 1978: group leader, deputy leader, chairman, secretary, chief whip, vice-chairman, deputy whip, assistant secretary/treasurer, four whips, three executive committee members.

In 1978 the elections for all fifteen positions were contested and in previous years most elections had also been contested. Fifteen members were a significant proportion of the Labour group: 34 per cent of the group as it was constituted from May 1976 to May 1978 (when there were forty-four members) and 37 per cent of the group as it was constituted from May 1978 to May 1979 (when there were forty members).

The leader was able to take advantage of the tradition of collective responsibility which applied to Labour group executive committee members. There was no reference to this in standing orders but it was a long-established practice. The group executive committee took votes and then every member of the executive committee had to support the majority decision in the full group meeting. This meant that the leader had merely to secure the support of seven other councillors in the executive committee to gain the fifteen-vote advantage in the full group meeting. This rarely proved difficult. However, the new leader's ability to secure majority support in the executive committee differed from that of the old leader. It had been extremely easy for the old leader to secure the support of the executive committee; his authority in the committee had been virtually unchallenged. Particularly during the first year or so in office the new leader's position was not unchallenged, and he was unable to count on the automatic support of the executive committee. The fairly undemanding task, under usual conditions, of securing support from seven other executive committee members was made easier by the practice of holding meetings of group officers prior to executive committee meetings. Usually these meetings involved the leader, deputy leader, group chairman, secretary and chief whip. Meetings of group officers were usually called to agree a common approach on potentially explosive issues such as those involving the direct labour building organisation.

Not only were executive committee members not allowed to vote against the decisions of the majority of the executive committee, they were also not supposed to speak against them. The usual reason advanced for this convention was that if it were not enforced then executive committee members would have two bites of the cherry. This was regarded as an inherently bad state of affairs. However, on one or two occasions under both leaders executive committee members did speak against executive committee decisions in a full group meeting. Nevertheless, with only one exception during the meeting I observed, no member – even those who spoke against particular decisions – actually voted against an executive committee recommendation.

Thus, the executive committee was a possible source of fifteen guaranteed votes which usually – but not inevitably – increased the power of the leader. It was not the only source of potentially guaranteed votes in the full group meetings. There was also the Policy and Resources Committee.

The tradition of collective responsibility also applied to the one-party Policy and Resources Committee. In 1976/7 there were fourteen members of the Policy and Resources Committee. Its membership overlapped substantially with that of the group executive committee but there were two Labour councillors who were members of the Policy and Resources Committee but not of the executive committee. The most important items which came before the group were presented to both the Policy and Resources Committee and the group executive committee beforehand. In such cases this gave the leadership a certain seventeen votes in the group meetings, that is, 39 per cent of the total votes. Often this meant that opposition on quite controversial issues was very small.

From time to time prior to May 1977 doubts were raised about whether or not collective responsibility applied to the Policy and Resources Committee. But during this period pressure was applied to particular individuals to support Policy and Resources Committee decisions in the full group. One councillor gave an example of this. He had been a communist in his younger days and still regarded himself as left wing. This meant that he normally opposed council house rent increases automatically. However, when, during 1976/7, a rent increase was under discussion, he voted in favour of it in the Labour group meeting. When asked about this after the group meeting he apologised but said in mitigation that he had been bound to support the decision of the Policy and Resources Committee (which was recommending the increase) regardless of his personal view.

Under the new leader the number of potentially guaranteed votes was even greater and the practice of collective responsibility for Policy and Resources Committee decisions was openly enforced. For example, the group chairman, who was a Policy and Resources Committee member, was opposed to the large rate increase initially proposed for 1978/9. She raised the issue at the group meeting and reported that she had been told that she must support the Policy and Resources Committee recommendation in the group, although she insisted on her right to speak freely.

For the municipal year 1977/8 the Policy and Resources Committee had been increased in size to fifteen. There was slightly less overlap between the executive committee and the Policy and Resources Committee memberships so that twenty councillors were members of one or the other. That is, 45 per cent of the group were either on the Policy and Resources Committee or the executive committee. Such a proportion was almost unbeatable. However, the proportion increased still further in 1978/9. For that year the Policy and Resources Committee had been increased in size from fifteen to seventeen. There were still fifteen members of the group executive committee and, altogether, twenty Labour councillors were members of either the Policy and Resources Committee or the group executive committee, that is, exactly 50 per cent of the Labour group (forty members in 1978/9).

In the case of the Policy and Resources Committee the principle of collective responsibility was more undesirable than in the case of the group executive committee, because Policy and Resources Committee members were appointed by the leader. This made it possible for him to stack the Policy and Resources Committee with his own supporters; this was not possible in the case of the executive committee because its membership was elected annually by the full group. Both leaders ensured that the majority of the Policy and Resources Committee were their supporters. This explains why Policy and Resources Committee was increased in size from fifteen to seventeen in 1978/9. The elections for the group executive committee had been fought between two factions: a well-organised faction supporting the old leader and a far less well-organised one supporting the new leader. The new leader managed to retain his own position but the overall results of the group elections that year meant that supporters of the former leader controlled the executive committee by eight votes to seven. This could have made it very difficult for the leader and so he ensured that he had a majority of support on the Policy and Resources Committee. He was unable to resist the claims of two councillors – who were both opposed to him – to become Policy and Resources Committee members, but overcame the problem by not removing two individuals who supported him.

A further problem was that the leadership was able to use one or the other or both of these bodies as it found appropriate. As one member of both the Policy and Resources Committee and the executive committee put it, 'The executive is used to bale out of the leadership when Policy and Resources Committee is divided'.[3]

PATRONAGE

One of the leader's tasks was to recommend to the group which Labour councillors should be members of which committees. More important, he also recommended which councillors should be chairmen and vice-chairmen of committees. According to Labour group standing orders, the membership of committees and chairmen and vice-chairmen were to be 'appointed by the leader in consultation with the group executive, subject to group approval'. In spite of this wording, up to May 1977 (under the old leader) the effective power to appoint chairmen and vice-chairmen lay in the hands of the leader alone. This is how one main committee chairman and executive committee member described how the system worked in 1974.

> With the present system councillors supposedly have some choice about the committees we get. The leader chooses and recommends and his recommendations are supposed to go to the executive committee. But it doesn't work. The last annual meeting was a fiasco and the executive had to be seen to be believed. The group secretary and the leader made a right mess, names were missed off by mistake . . . The group wanted the executive committee to have a say in the nominations but it didn't. The executive had no chance to question the leader's nominations . . . At present anything can happen.

He went on to mention the system that had previously existed. Candidates for office had to be nominated at the group annual general meeting and a vote was then taken. 'It didn't work perfectly', he said, 'but in the main we got good chairmen; nine out of ten were good chairmen. Of course you could end up with nothing or with something you didn't want. But it was better than the leader bestowing patronage.'

Other studies have revealed that a leader's power of patronage sometimes became an end in itself. Baxter gives an account of John Braddock's domination of Liverpool's Labour group after 1945. A

basic characteristic of the Braddock years was a lack of policy innova-
tion. The main consequence of his dominance was the channelling of
patronage into favoured quarters (Baxter, 1972). Similarly, Sam Boyce
dominated the Labour group in West Ham from the mid-1930s until the
early 1960s. The group under his sway 'largely avoided political principles as a topic of debate and concentrated on enjoying patronage,
power and prestige' (Gyford, 1976, p. 74). In Newcastle the chief
beneficiaries of the old leader's powers of patronage were friends and
loyal retainers. During the new leader's period of office patronage was
used in a different way. He certainly used it to secure his own continued
tenure of the office of leader; but he also sought to promote individuals
of ability – who seemed to many group members to have been deliberately ignored under the previous regime – to positions of influence and
responsibility. Offices were distributed in the tacit expectation of
support in future elections for group leader, but also with a view to
getting particular jobs carried out effectively.

Although the new leader also had considerable power over appointments, his power was not unchallenged. The group executive committee
took a more critical look at his nominations for the municipal year
1977/8 than they had in previous years, and they changed one (the vice-
chairmanship of a main committee).

In the following year there was organised opposition to the leader's
initial proposals and he was unable to get sufficient support in the
executive committee. He then produced modified proposals to which
there was also strong organised opposition. The campaign against his
proposals was led by the former leader and his wife. It took an
extremely personal form which the leader felt it necessary to comment
on at the full group meeting. He described how one hostile executive
committee member had said of him, in an executive committee meeting,
'Leader? He couldn't lead a troop of girl guides.' The leader insisted
that the vote of the executive committee was not binding on him since
standing orders only required him to consult the executive. The vote of
the executive committee (against his proposals) was therefore, he
argued, only advisory. The leader then announced that he would resign
if his nominations were not approved. As a result they were agreed *en
bloc* by the full group without any dissenting votes being cast. Thus, in
spite of the challenge, he had successfully distributed the overwhelming
majority of the most prized official positions.

What effect did the leader's ability to distribute positions of power
have on the behaviour of Labour councillors, particularly those who
were favoured by the leader? Did it increase the hold of the leader over
the group? My own observations indicate that it did. No doubt both
leaders would deny that their use of patronage affected the voting
behaviour of Labour councillors and indeed few councillors would
admit to being bought off. However, one subcommittee chairman

(then a member of both the group executive committee and the Policy and Resources Committee) was honest enough to describe how it affected him to a district Labour party working group in 1974.

> I admit to a certain amount of being bought off by being made chairman of [a named] subcommittee. In return you are probably less critical of the person dispensing the patronage. It's not just that though. You need to specialise in order to get things done. Previously I was critical about many aspects of policy. Now I tend not to be critical of everything. It's the only way it can be done though. I want to get a sensible [named] policy carried out and it's the only way to do it.

As long as he was left to get on with his own specialism, he refrained from challenging the leader on other subjects during group meetings. This was how the majority of the chairmen and vice-chairmen acted. It worked this way because there was not a single centre of power in the local authority but rather several specialised centres of power. Each of these centres of power was normally dominated by the appropriate chairman and chief officer. Most councillors were content to be able to dominate a particular specialism, that is, a committee or subcommittee, and in doing so struck a tacit bargain with the leader.

How many Labour councillors did this patronage involve in the municipal years 1976/7, 1977/8 and 1978/9? (The 1976/7 distribution took place during the period in office of the old leader and the other two under the new leader.)

Municipal Year 1976/7

During this year there were ten main committees and thirty subcommittees or similar bodies which all met regularly (normally monthly). There were also twelve subcommittees or similar bodies meeting as and when required and five advisory bodies, four of which met regularly and one, as and when required. Recommended appointments, including those of chairman and vice-chairman, were submitted to the group by the leader for all but one of these bodies (an advisory committee). The group had the power to reject any of the appointments recommended by him but agreed them all *en masse*. (Appointments were also made to sixty-three outside bodies.)

There were forty-four Labour councillors. Of these, seventeen were made chairmen or vice-chairmen of the ten main committees. A further sixteen Labour councillors were made chairmen or vice-chairmen of the subcommittees and advisory bodies. Thus, thirty-three Labour councillors were given official positions. That is, exactly 75 per cent of the Labour group held highly prized positions, all of which had been given them by the leader.

Municipal Year 1977/8

In this year there were eleven main committees, thirty-four sub-committees or similar bodies meeting regularly, thirteen subcommittees or similar bodies meeting as and when required and five advisory bodies. (Appointments were also made to sixty-eight outside bodies.) The Labour group were asked to agree the appointments to eleven main committees, thirty-four subcommittees or other bodies which met regularly, thirteen subcommittees or other bodies meeting as and when required and four advisory bodies. This included the appointment of the chairmen and vice-chairmen of each body.

Twenty-one councillors were made either chairmen or vice-chairmen of the eleven main committees and fifteen additional councillors were made chairmen or vice-chairmen of the subcommittees which met regularly. The other appointments went to councillors already holding one of the above positions. Thus thirty-six Labour councillors were given positions during the year 1977/8. That is, 82 per cent of the group were awarded official positions by the leader. Only eight councillors remained without any office. All but one of the chairmanships and vice-chairmanships had been allocated by the leader. The full group had agreed *en masse* all the appointments recommended to it by the leader and the executive committee.

Municipal Year 1978/9

The Labour group made appointments to eleven main committees, forty-one subcommittees or other bodies meeting regularly, nineteen subcommittees or other bodies meeting as and when required and four advisory bodies, including the appointment of chairmen and vice-chairmen. (In addition representatives on sixty-seven outside bodies were appointed.)

Twenty-one Labour councillors were made chairmen or vice-chairmen of the eleven main committees and a further fourteen Labour councillors were made chairmen or vice-chairmen of the sub-committees or other bodies. That is, thirty-five councillors out of a total of forty (four seats had been lost in the elections of May 1978) were allocated highly valued positions by the leader; they formed 87 per cent of the group. In addition, the membership of the high-powered Policy and Resources Committee was increased from fifteen to seventeen.

The great power of prime ministers which results from being able to allocate offices to members of their own party is often criticised. However, not even a prime minister is able to offer positions in the government to 87 per cent of his or her party. See Table 6.1.

The issue of patronage was frequently raised by the local party and was a further source of tension between it and the group. The group was asked to vote on the issue at its 1975 annual general meeting and voted against change. A year later, at its meeting in May 1976, the district

Table 6.1 *Patronage, 1976/7, 1977/8, 1978/9*

	1976/7		1977/8		1978/9	
	Number of bodies submitted to group	Potential positions to be allocated	Number of bodies submitted to group	Potential positions to be allocated	Number of bodies submitted to group	Potential positions to be allocated
Main committees	10	20	11	22	11	22
Subcommittees meeting regularly*	30	60	34	68	41	82
Subcommittees meeting as and when required*	12	24	13	26	19	38
Advisory bodies	4	8	4	8	4	8
TOTAL	56	112	62	124	75	150

*Includes other bodies similar to subcommittees.

Labour party resolved the following: 'That a motion should once again be submitted to the annual meeting of the Labour group in the following terms: "This group resolves that the chairmen and vice-chairmen of council committees shall be elected by the group." ' The wording of the minute reflects the exasperation of the district Labour party, but the suggestion was again rejected by the group with only four votes in favour. The issue was raised again by the district Labour party in December 1977 but on this occasion the group rejected the proposal without even taking a vote. Too many Labour councillors feared that they would be worse off under a new arrangement.

CONCLUSION

The leader in Newcastle upon Tyne was without question a very powerful figure. Even when his position was being challenged his power remained substantial. Was this power, however, completely without limits? Robert Michels saw the power of leaders over followers as being almost unlimited. But, as Robert McKenzie argued, they could not ignore their followers completely. Labour party members have been traditionally committed to a number of goals in the local government context, such as not selling council houses and not making members of the council's workforce compulsorily redundant. The group leadership in Newcastle certainly had to take account of such values. Indeed the former leader, who normally experienced no difficulty in securing majority support in the group executive committee, found that he was unable to get the executive's agreement to a proposal to sell limited categories of council houses to sitting tenants. At its meeting in July 1976 the scheme he was putting forward was suspended by the group on the recommendation of the executive. (The scheme was suspended rather than rejected. This helped to give the impression that the decision was not really a defeat for the leader, and that all that was at stake was a question of timing.) Thus traditional party values provided limits to the power wielded by the group leader. However, as long as Newcastle's leader did not stray outside those traditional limits, he enjoyed extensive freedom of action.

NOTES: CHAPTER 6

1 *Newcastle Journal*, 10 May 1977.
2 *Newcastle Journal*, 31 January 1975.
3 Interviewed in November 1974.

Chapter 7

GROUP DISCIPLINE

I have referred to party discipline on several occasions earlier in this book. It is now time to examine the issue more systematically. I shall consider first the constitutional framework and then what the position was in Newcastle upon Tyne.

THE CONSTITUTIONAL FRAMEWORK

The group's standing orders were based on the model standing orders issued by Transport House, with only minor differences. These model rules are contained in a document entitled *Labour Groups on Local Authorities* along with some official advice about their interpretation (Labour Party, 1975c). This advice has recently been subject to slight modification following the report of a special committee of the National Executive Committee (Labour Party, 1975a), although the standing orders themselves were not modified. Newcastle Labour group standing orders on the subject of party discipline read as follows:

(a) Individual members of the group shall not submit or move resolutions, notices of motion or amendments unless such resolutions, notices of motion or amendments have first been submitted to and received the approval of the group leader or his deputy.

(b) Individual members of the Labour group may ask questions at meetings of the council, provided the tendency of such questions is not likely to be in conflict with the policy of the Labour group, and the group leader has been consulted.

(c) Members of the Labour group are expected not to speak or vote at meetings of the council in opposition to the decisions of the Labour group, unless the group have decided to leave the matter in question to a free vote. (Where matters of conscience arise, e.g. religion, temperance, etc., individual members of the Labour group may abstain from voting, provided they first raise the matter at the group meeting in order to ascertain the feeling of the group)....

It is hereby declared that acceptance of these standing orders is a condition of membership of the Labour group on the council. It shall be competent for the group whip to be withdrawn from any member who violates these standing orders, such action to be determined upon by the Labour group upon due notice and after consultation with the executive committee of the district Labour party, subject to appeal, within fourteen days, to the National Executive Committee of the Labour Party, whose decision shall be final.

At its annual general meeting the group elected a chief whip, a deputy whip and four assistant whips to enforce these rules. On the committees it was the job of each vice-chairman to act as committee whip. There was, therefore, no shortage of individuals whose job it was to ensure adherence to standing orders.

THE SCOPE OF PARTY DISCIPLINE

There is a very strong tradition in the Labour Party (and the trade unions) of accepting the majority decision. This was put in an extreme form by Hugh Scanlon: 'Liberty is not licence. *Liberty*, in my view, *is conforming to majority opinion.* It is not doing what you want.' (my italics)[1]

The doctrine that the minority must submit to the majority is normally justified by its adherents on the ground that all decisions are preceded by full and fair discussion. It is a point of view which predates the existence of the Labour Party. This is what Schnadhorst, a leading organiser of the Birmingham Liberal caucus, said on the subject in a speech to the Cambridge Reform Club in 1878.

No opinion, creed, or position shut a man out. So long as the minority are content to submit to the majority, no matter what opinion upon any particular question a man held, he was not excluded. Free discussion was granted to all so long as perfect loyalty to each other existed.
(Ostrogorski, 1902, Vol. I, p. 183)

However, it is a view which has many supporters within the Labour Party. W. J. M. MacKenzie described the Labour Party as 'governed under the myth of majority decision after full and free discussion'. The party could not deny this myth without denying its own existence, he said (MacKenzie, 1955, p. 158). The view that party democracy is majority rule also has its critics. For example, Aneurin Bevan argued, according to Michael Foot, that 'the majority theory that the first duty of Members of Parliament was to uphold the policies of Labour ministers with binding decisions at secret party meetings was a recipe for making Parliament a rubber stamp'. 'In this fashion', Bevan wrote, 'political parties become the enemies of parliamentary democracy . . . It is better that ministers should be embarrassed than that Parliament should die . . . Party discipline should end where it impinges on parliamentary liberty' (Foot, 1975, pp. 330–1).

Party discipline can be consistent with democratic values. One reason that political parties in local government have internal discipline is to ensure that the wishes of citizens, as expressed at election time, will be acted upon. In recent years this notion of the electoral mandate has assumed greater prominence in British politics. It increases the accountability of governments to the electorate when a party's manifesto policies have been subjected to full publicity and discussion. If and where this applies, a political party is justified in imposing strong discipline on its members to secure agreement to its election pledges.

It applies, however, to only a limited number of policies in local government. Most policies have been initiated in private, and in so far as they have been subject to examination of any kind, it has also taken place in private. To employ party discipline to force through proposals of this provenance is inconsistent with the democratic values of those who would agree with Bevan's conception of democracy.

Bulpitt found that Labour groups in the north west of England varied in their interpretation of standing orders. The Labour groups in Sale, Salford, Manchester and Rochdale had adopted a loose interpretation of standing orders, that is, no matter coming before the group was regarded as a whip issue unless the group resolved that it should be so regarded. The Middleton and Macclesfield Labour groups adopted a strict interpretation of standing orders: every matter coming before the group was a whip issue unless the group decided to the contrary (Bulpitt, 1967, p. 100).

In Newcastle a strict interpretation of standing orders was applied. Every decision of the full group and of committee pre-meetings – whether or not it had formed part of an election manifesto – was automatically regarded as a whip issue. There was only one free vote during the period of this study; it occurred in 1978 when Labour councillors were allowed a free vote on whether or not the council should organise its own lottery.

Labour councillors were asked about their attitude to party discipline during interview. They were asked to say whether they agreed or disagreed with the following statements:

A councillor should *always* abide by the decision of his party group or pre-meeting in the council or council committee meetings. (question 76)

A councillor should be willing to vote against the policy of his party group or pre-meeting in the council or its committees if he believes it to be harmful or unjust. (question 77)

All forty Labour councillors answered question 76. Thirty (75 per cent) agreed and ten (25 per cent) disagreed with the statement. Of the thirty who agreed, twenty (50 per cent of the total) did so strongly. Of the ten who disagreed, only two (5 per cent of the total) did so strongly. Respondents who agreed strongly had this to say:

> That's the democratic world we live in. If you're beat that's it. I don't even think you should abstain. No. An abstention is a vote for the other side. (a backbench councillor)

> You couldn't function any other way. It would be futile. You'd be up the creek. (a main committee chairman)

> I'm one hundred per cent in favour. I've never voted or spoken against a party decision. (a main committee vice-chairman)

Four of the Labour councillors who said they agreed with the statement, but not strongly, expressed some reluctance in agreeing. For example, a subcommittee chairman said, 'I'm a loyalist so I must agree. I've often had to put the case for my constituency Labour party and then to abide by the majority decision. You've got to have loyalty. I don't believe the party is right irrespective of anything else but it's the only one we've got so you shouldn't shake it overmuch. You should argue inside the party.' And a backbench councillor said, 'You've got to have discipline but it goes too far. You've always got to support party policy. You can't speak up to represent the people. You've got to toe the line too much. A lot of the other councillors feel you should keep your mouth shut.'

A strong feeling among Labour councillors was the need to keep a united front. For example:

> I have to say yes, reluctantly. You need to keep a united front in the face of the opposition. It's one of those unfortunate things. You need to keep out the opposition. (a main committee vice-chairman)

The above councillor thought that this necessitated obedience to the views of the majority. Indeed this was the official view of the group. For example, group members received the following letter from the chief whip in February 1976:

Dear Colleague,

... with a tough time up to the May elections, it is essential that we are seen to be a united party.

Members are reminded that pre-meetings are held in order to agree a common approach and differences of opinion should not be aired in committee. Such action plays into the hands of the Tories and the press.

If a member thinks that a committee is on the wrong tack and feels unable to accept the chairman's advice, he has recourse to the leader and ultimately the group.

Please act responsibly,

Yours sincerely,

(signed)

chief whip

Many of the Labour councillors who disagreed, but not strongly, expressed agreement with the principle of always supporting group decisions but referred to the right to abstain contained in standing orders. They only disagreed in this somewhat restricted sense – restricted since the right to abstain was in practice very narrowly defined and hardly ever taken advantage of. For example:

> I agree ninety-seven per cent. But there will always be times when someone feels so strongly about an issue they don't want to support it. They could abstain. (a main committee chairman)

Few expressed clear disagreement in the way that this main committee vice-chairman did. He said, 'I disagree. There is the conscience clause. There must be a let out.'

Question 77 was asked in order to measure the strength of feeling in favour of supporting party discipline. Answers to question 77 correlate very closely with those for question 76. Forty Labour councillors answered question 77. Twelve (30 per cent) agreed with the statement and twenty-eight (70 per cent) disagreed. Of the twelve who agreed, only three (7 per cent of the total) did so strongly. Of the twenty-eight who disagreed, ten (25 per cent of the total) did so strongly.

One of the backbench councillors who strongly agreed with the statement referred to the leadership. He said, 'I strongly agree. Acceptance of everything the leader says is just stupid.'

One backbench councillor agreed with the statement but only in circumstances where the group's policy was opposed to that of the district Labour party. Another councillor, a main committee vice-chairman, said he agreed 'on moral issues anyway'.

Two councillors who agreed with the statement referred to the disciplinary consequences of voting against group decisions; one of them said that in practice this deterred him from ever doing so. Both were vice-chairmen of main committees. The first said, 'I agree. You would then have to take the rap.' The second said, 'There are things I would like to vote against but to put it bluntly I wouldn't have the moral courage, if that's the right word. You would incur the wrath of the whip. So having said I agree, I wouldn't do it. I wouldn't have the moral courage.'

The following councillors were typical of those who strongly disagreed with the statement.

You should not vote against group decisions. You can speak but when it comes to voting, the group has reached a decision. You've got to accept it. You can't vote with the Tories. You should never do that. (a backbench councillor)

You must abide by the group decision. I've done it. I've strongly disagreed but I've always abided by the decision. (a main committee vice-chairman)

Councillors who disagreed but not strongly, had this to say:

You can vote against a policy at the group. That's the democratic method. I believe in free speech but only in the pre-meetings. Not when the Tories are there. At group you've got the chance to put your case. If you're defeated you've got to accept it. (a subcommittee chairman)

You shouldn't vote against. If there is a pre-meeting, at the end of the road you should abide by the majority decision. If a chairman went through all the rigmarole of a pre-meeting then he would want you to abide by the decision. (a main committee vice-chairman)

TAKING DISCIPLINARY ACTION

Bulpitt found that disciplinary action by Labour groups 'never assumed any great importance' (Bulpitt, 1967, pp. 99–100). During the period of his study of six authorities (May 1955–May 1960) he found that only two Labour councillors had the whip withdrawn (p. 101).

Newton found that, at Birmingham, 'full-scale disciplinary matters' were rare. Indeed he says, 'they are so rare that their memory as *causes célèbres* lingers on for years afterwards' (1976b, p. 262). In Birmingham, there had only been one case since 1950: in 1963 eight Labour councillors had the whip withdrawn and an attempt was made to expel them from the party altogether (pp. 259–60; eventually the eight councillors were readmitted to the group).

In Newcastle there was very strong support for the view that decisions of the majority must be supported. As a result it was very rare for Labour group members to defy a decision of the full group or of a committee pre-meeting. During the period of this study no Labour councillor voted against a group decision in a full council meeting. However there were two other cases of Labour councillors voting against the party line. One subcommittee chairman reported during interview that he had voted against a main committee pre-meeting decision in support of the policy of local trade unions on that subject. The issue had been whether or not warehousing was an appropriate use for land formerly used by a nationally known heavy engineering firm. He felt it was legitimate for him to have supported the policy of his trade union instead of the pre-meeting decision. No action had been taken against him. The second case concerned a Labour councillor who voted against a decision of the full Labour group at a subcommittee meeting. The councillor was interviewed by the chief whip, but again, no action was taken.

Jones found that the Wolverhampton Labour group adopted a similar approach: 'The common practice was for a deviant to be reported to the group, when an explanation was invited and was usually accepted.' The Wolverhampton Labour group did not, however, have the same commitment to supporting the majority view as at Newcastle. Jones reports that there were numerous cases where Labour councillors voted against group decisions (1969, p. 183). Newton found that the normal practice at Birmingham was for breaches of discipline to be dealt with by issuing a warning to any rebels. He records how four Labour councillors rebelled against the group on one occasion: 'some hard words' were 'delivered by group members and officers' but none of the rebels had any formal action taken against them (1976b, p. 259). This was also how disciplinary questions were handled in Newcastle: disciplinary action normally took the form of issuing warnings to individual councillors, or seeking assurances from them about their future conduct. During the period of this study no Labour councillors had the whip withdrawn.

I consider now two examples of disciplinary action which were taken during the period prior to May 1977. The first concerned two councillors who were invited by the shop stewards' committee of the direct labour building organisation to a meeting to discuss the grievances of the workforce. The DLBO had been making heavy losses and the workforce was being blamed, certainly in the press, and either directly or by implication in official council reports. The shop stewards felt that this was unfair and argued that, among other things, the low productivity of the organisation was due not to their failure to work hard enough, but to massive overmanning (manning levels are not of course under the control of shop stewards or the workforce itself) on some sites. (Their view about the importance of overmanning turned out to be correct.) The two councillors attended the meeting and as a result they were ordered by the chief whip to appear before the whips to explain their conduct. The chief whip set out the reasons for this meeting in the following letter:

> Dear [name],
>   . . . it was reported . . . that you visited a Direct Labour building site and talked to the workers and or their shop stewards to discuss their grievances. Apparently these alleged talks infringe on existing industrial relations machinery. The men's full-time union officials have made a complaint. It is about this matter that we wish to see you. . . .
>
> Yours fraternally,
(signed)
chief whip

It turned out that a full-time official of one of the several unions involved had criticised the presence of the two councillors at the site meeting. (He had been there himself but had said nothing at the time.) Much was made of this complaint by the whips. Indeed it was the main argument put forward to support the view that councillors should not speak to shop stewards and discuss their grievances.

In the author's judgement this was not the real reason for the attempt to discipline the two councillors. The DLBO was being heavily criticised in the press because of the enormous losses it was incurring. The response of many Labour councillors when being criticised was, as the interviews show, to resort to secrecy (see below, p. 167 – 8). This was the attitude of the leadership. If they succeeded in preventing the two Labour councillors from meeting the shop stewards again, the consequence would have been to limit the opportunities for potential critics to gain first-hand knowledge of the problems being experienced by the DLBO. Clearly to do this limited their potential effectiveness.

I recount in some detail what was said at the meeting with the whips and at a subsequent group meeting. I have done so in order to convey more accurately the character of the relationship between fellow members of the group in such circumstances.

The two councillors,[2] the chief whip, the deputy whip and two of the four assistant whips were present at the disciplinary meeting:

Chief whip: The group executive committee has instructed us to see you because it has been reported that you have been on a direct labour building site and talked to workers and shop stewards and listened to their grievances and possibly intimated to them that you were in sympathy with them. [A named full-time trade union official] has complained that you were subverting the existing machinery and making a mockery of it. You were doing the job of the full-time trade union officials. What would happen if all councillors spoke to all the unions? How many unions are there? There would be turmoil.

Councillor A: What turmoil was there in this case?

Chief whip: Well, none.

Councillor A: We went to the meeting, certainly; but does this infringe anything? The answer is no.

Councillor B: Let me explain how this came about from the beginning. The district Labour party has set up a housing policy working group of which I am chairman. We discuss various aspects of housing policy. When we discussed direct labour we invited the shop stewards to come along and they sent six out of twelve of their committee. They had felt that the meeting was worthwhile and that they had received a fair hearing. After this they asked [Councillor A] and I a couple of weeks later to a meeting to be held in their lunch break. We were invited the night before the meeting, in our capacity as members of the housing policy working group — myself as chairman and [Councillor A] as the member with special responsibility for the item in question . . . So we were there as members of the district Labour party housing policy working group, although obviously they knew we were councillors.

Assistant whip A: They were making use of you, you mean.

Councillor B: We were not acting in an executive capacity and we had explained this to them. The only way that meeting with us could be effective was in the long term through the district Labour party . . . [Councillor A] and I both argued that we felt the majority of the Labour group would be sympathetic and be willing to hear their case. We suggested that they should contact the Labour group direct since they felt that they could not get a fair hearing from council officers. A major complaint was that the shop stewards only met council officers and not elected representatives . . . We suggested that the solution to this was for them to meet with the chairman and vice-chairman of the Direct Labour Organisation Committee.

Deputy whip: If they had resentments they should have used the full-time officials. If all the unions refused to use their full-time officials there would be turmoil.

Councillor A: We were not there instead of the full-time official but as well as him, and we at no time claimed to be able to intervene, even if we wanted to, which we did not. We were just there to give such advice as we were asked for. The full-time official was introduced to us outside and made no attempt to say that we should not go in.

Deputy whip: OK, they came to a meeting of your working group – we'll accept that, but it's highly dangerous to go on a site and listen to complaints.

Councillor A: So if they phone up and ask us to go to a meeting – as they did – we have to say no?

Assistant whip A: Yes. If the district Labour party wants information they have members who are not councillors who they could ask to get it.

Chief whip (interrupting): Anyway will you give an assurance that the next time you will tell them to meet either the group secretary or the chairman or vice-chairman of the Direct Labour Organisation Committee?

Councillor A: We told them to contact the group – isn't that the right and sensible thing to do?

Chief whip: Will you agree in future to refer the men to the proper channels?

Councillor A: We did this time.

Assistant whip B: The executive committee felt that industrial relations had been interfered with. Will you give an assurance that you will not interfere again? The executive committee can then decide.

Councillor B: But we have not interfered.

Councillor A: Clearly 'interfere' is a critical word here. What exactly does it mean? One question is, did we or did we not follow correct procedure?

Assistant whip A: If you went to the meeting, no.

Chief whip: If you told them to contact the group or the direct labour committee you have carried out correct procedure.

Deputy whip: You were there as councillors. Let's face it, being a councillor can open a few doors.

Councillor A: It can also close quite a few.

Chief whip: Anyway, let's not start all over again. We'll report to the executive committee.

The matter was raised at the end of the group meeting in May 1976. The chief whip reported on the meeting between the two councillors and the whips, and asked the group to adopt a resolution that councillors should not in future interfere in industrial relations. A debate followed, during which the chairman of the Direct Labour Organisation Committee, speaking in a bitter tone, said: 'I was not told by these councillors that they had attended this meeting . . . The work of the organisation could be undermined by a sinister fifth column. We don't want people stirring up trouble.'

The discussion became heated at one point but the meeting ended by agreeing to the resolution put forward by the chief whip.

I turn now to the second example. During the row over the DLBO the group chairman – who was also district Labour party chairman and vice-chairman of the Direct Labour Organisation Committee – was subject to disciplinary action. At a meeting of the council's Works Committee (a joint committee on which councillors and all the main trade unions were represented) the problem of the DLBO had been discussed. The group chairman had asked for it to be recorded in the minutes of the Works Committee meeting that the difficulties of the DLBO had been caused by decisions of the Labour group. The leadership objected to this criticism and at a meeting of the group executive committee it was agreed that the chief whip and deputy whip should speak to the group chairman about his remarks. However, before he was interviewed by the whips the group chairman insisted that the district Labour party secretary and the Labour party regional organiser should be in attendance along with representatives of the appropriate trade unions. He knew that he could rely upon support from the district Labour party and the trade unions and that this made it difficult for the whips to act against him.

The whips warned him about his future conduct but no action was taken against him.

In both cases the support of the district Labour party had been important. Without district Labour party support it was difficult for the Labour group to make any sanctions tell. Newton found that the Birmingham Labour group were similarly constrained. In the case of the eight rebels, referred to above, a major factor in their favour had been the support which the Birmingham borough Labour party had given them. Newton concludes that 'rebels are not powerless by any means, especially when they are backed by their ward organisations, by the borough party, by the ambiguity surrounding Clause XII (i),[3] and the need for the group to present a united front to the public at election time' (1976b, p. 261).

These factors also applied in Newcastle. The ward parties were important because they actually selected individual candidates. As long as a councillor kept the support of his ward he was in a strong position. (It was rare for wards to sack their councillor, but it happened in two cases during the period of this study.) The district (or borough) Labour party was important because it controlled the approved list of candidates (the panel) from which wards had to select their own candidate. As long as a councillor had sufficient support in the district Labour party the group could do very little to him. In Newcastle the district Labour party was critical of group policy towards the DLBO and the three councillors, in the two examples above, knew that they could count on support from that quarter. (Indeed, shortly after the above events, the district Labour party issued a warning to Labour councillors that it would remove them from the panel if they failed to support district Labour party policy.)

THE FREEDOM OF THE INDIVIDUAL COUNCILLOR

Over the years Transport House has issued a series of appeals for Labour groups to relax control of the actions of individual members. The report of the special committee on the conduct of the Labour Party in local government had the following to say on the subject of party discipline:

> Although in August 1973 the National Executive Committee advised Labour groups to review the operation of their standing orders so that there might be less rigidity on matters which do not involve the election programme or policy issues, we found that some groups do not permit any flexibility whatever . . . We are strongly of the opinion that Labour groups should permit the maximum flexibility and allow members to raise local issues which are not contrary to party or group policy. This is provided that the matter is first raised at a group meeting, or in a case of urgency, with the group leader.
> (Labour Party, 1975a, p. 8)

As a result of this special committee report a change was made in the official advice given to Labour groups. The following sentence was added to the section entitled 'Standing orders of Labour groups'.

> The maximum amount of flexibility should be allowed for members to raise local issues that are not contrary to party or group policy and, whenever possible, submitting such non-political matters to a free vote.
> (Labour Party, 1975c)

Bulpitt describes how the Middleton Labour group disciplined the deputy chairman of their Highways Committee because 'in September 1959 he publicly congratulated the town's Junior Road Safety Committee on their part in persuading the Minister of Transport to agree to the siting of some new traffic lights'. The whip was withdrawn and he was sacked as deputy chairman of the Highways Committee (1967, p. 26). What was the attitude of the Newcastle Labour group towards the freedom of the individual councillor? During the period before reorganisation Jon Gower Davies recounts how, in Newcastle upon Tyne, he was prevented from presenting a petition to the council which criticised the council's motorway policy. He was told that it was against group policy to criticise the motorway. He had to present the petition without saying, in council, what it was about (Davies, 1972a, p. 113). After reorganisation similar pressures to refrain from criticism were put on councillors. In an interview with a district Labour party working group in 1974 the then chief whip reported that he had been told to discipline a subcommittee chairman, coincidentally also Jon Gower Davies, merely for opposing the views of the then leader. He said, 'When Jon Davies challenged the leader on one thing I, as chief whip, had to carpet him – just for challenging the leader'. I consider now an example of the Newcastle Labour group's attitude towards the freedom of the individual councillor. This is followed by an account of what Labour councillors said on the subject during interview.

Freedom to Express an Opinion in Public
The freedom of the individual member became an issue in January and February 1977 when a Newcastle Labour councillor appeared on a television programme. It was discussed at the Labour group meeting held in January 1977. At that meeting the chief whip said:

> We can't all act as individuals. We must act as a group. You can't go around willy-nilly making statements, writing letters and so on or we would have chaos. We can't all act as individuals. This only applies to controversial statements. For example if [a named councillor] was talking about Newcastle United [as he had done in the last few days] that would not need clearing. It is a matter of courtesy, common sense and decency to tell your colleagues when a controversial statement is to be made.

(The example he gave concerned a statement made about the local football team which had had nothing whatever to do with the council.)

After the meeting group members received the following letter from the chief whip:

> Dear Colleague,
>   At the Labour group meeting held on Monday 31 January 1977 it was agreed . . . that I should write to individual group members informing them not to ignore the group's previous decision when writing letters and articles to the press and making statements on radio and television.
>   The decision was that individual group members should not go to the media with, what could be, controversial statements without first clearing them with the relevant committee chairman and/or the Labour group leader.
>   Yours sincerely,
>   (signed)
>   *chief whip*

The issue was raised at the next group meeting by a councillor who argued that the chief whip's letter went too far. It implied that a chairman or the leader could prevent a councillor from appearing on television or making a statement. He asked what the letter meant. Did members have to inform the chairman or leader purely as a matter of information or did the chairman or leader have the right to refuse? This all-important distinction was obscured by the chief whip who, in his reply, simply laid heavy stress on courtesy. But the clear intention of the leadership had been to restrict the ability of Labour councillors to make public statements.

The Interviews
The Labour councillor's views about the freedom of the individual member were tested by question 84. This question is based on a statement in the report of the special committee of the National Executive Committee (Labour Party, 1975a, p. 8, para. 6.6). Respondents were asked to say whether they agreed or disagreed with the following statement:

> Party groups should allow their members to raise in council or in its committees any issue which is not contrary to group policy.

Thirty-four Labour councillors answered this question. Twenty-three (68 per cent) agreed and eleven (32 per cent) disagreed with the statement. Of the twenty-three who agreed, nine (26 per cent of the total) did so strongly and of the eleven who disagreed, only three (9 per cent of the total) did so strongly.

Some of the Labour councillors who said they agreed with the statement qualified their agreement. For example:

> I agree as long as you tell the chairman first. It's only right to tell the chairman it will be raised, to give him due warning.
> (a subcommittee chairman)

> Such issues could be raised in group anyway. I would only be in favour if a councillor raised the issue in group first. It would have to be discussed in group first.
> (a main committee vice-chairman)

Of the councillors who said they disagreed with the statement one, a subcommittee chairman, said she would go along with such a policy in a limited number of cases only:

> I agree only if the issue was of very great importance. You could have awkward people when we are in opposition raising ever so many

issues. You could allow for emergency questions or resolutions. Councillors shouldn't raise *any* issue.

One subcommittee chairman felt that the permission of the relevant chairman or of the leader should be required. He said, 'I disagree. Who decides what is group policy? It could be a borderline case. No, it should be with the permission of the chairman or the leader.'

One backbench councillor was worried that the party might be embarrassed if councillors had the freedom to raised such issues. He said, 'No, councillors should not raise issues which are not contrary to group policy because it might embarrass the party. It depends on the issue.' Finally one subcommittee vice-chairman said, 'I can't see the need for it. If it's not against policy why raise it?'

NOTES: CHAPTER 7

1 Hugh Scanlon, quoted in the *Listener*, 11 August 1977.
2 The author is referred to as Councillor B.
3 The clause of the Birmingham borough Labour party standing orders (based on Transport House model standing orders) which laid down its right to determine Labour Party election policy locally.

Part Three

THE COMMITTEES

Chapter 8

THE DISTRIBUTION OF POWER IN THE
COMMITTEE PROCESS

In some places the informal and unofficial, but very real power,
which chairmen may acquire has resulted in an unwholesome
Caesarism.
(J. H. Warren, 1946, p. 106)

A member with a strong personality, who has served for some years
as chairman of a committee, who stands high in the inner councils
of his party, and is trusted by its rank and file, can do almost what
he likes both with his committee and the council . . . If the chairman
is not an able man (which happens quite frequently) he finds it
easier to get his business through by using his prestige than by openness and argument. The result is that the council knows nothing
about the work of his committee and loses all control.
(Lord Simon of Wythenshawe, writing about Manchester City
Council, 1926, p. 61)

Labour councillors were asked a series of detailed questions (questions
31–42) about one committee of which they were (with one exception)
backbench members. They were asked to describe the power of the
principal actors by reference to their effort and efficacy. There were
questions about the relevant chief officer, the chairman, the individual
backbencher in the majority party and the individual backbencher in
the opposition party (or parties). The following results were obtained.

Seventy-seven per cent of Labour councillors who answered the question about committee chairmen thought that the chairman under discussion had a big effect on the decisions of his or her committee. Nearly
64 per cent of respondents thought that the chief officer in question had
a big effect on the decisions of the committee. The overwhelming
majority of Labour councillors (90 per cent) thought that the majority
party backbencher would have to put in either a lot of effort or (in two
cases) that it would make no difference how much effort they put in. A
smaller but still very large majority (71 per cent) thought that backbenchers had either little effect or no effect. Every respondent thought

that opposition party backbenchers had to put in a lot of effort, and twenty-nine respondents (91 per cent) thought they had little or no effect. See Table 8.1.

Table 8.1　*A Comparison of the Relative Power of Chairmen, Chief Officers, and Majority and Opposition Party Back-benchers*

	EFFORT				EFFECT			
			Little/	Would make no				
	Lot	Some	None	Difference	Big	Some	Little	None
	%	%	%	%	%	%	%	%
Chairman	28	31	41	—	77	20	—	3
Chief officer	40	27	33	—	64	31	6	—
Majority party backbencher	85	8	3	5	3	26	56	15
Opposition party backbencher	74	—	—	26	—	9	25	66

These data clearly suggest that committee chairmen were pre-eminent. This view is also supported by the results obtained from two further questions, numbers 52 and 53. Twenty-nine chairmen and vice-chairmen of main committees and subcommittees[1] were asked to state how much power they thought they had and whether or not they were satisfied with it.

The main committee chairmen's answers are set out in Tables 8.2 and 8.3.

Table 8.2　*Chairmen of Main Committees: Perception of Own Influence*

Influence:	Lot	6	55%
	Some	3	27%
	Little	2	18%
	None	0	—
	Total	11	100%

Table 8.3 *Chairmen of Main Committees: Satisfaction with Own Influence*

Attitude:			
	Very Satisfied	5	45%
	Satisfied	4	36%
	Qualified answer	1	9%
	Dissatisfied	1	9%
	Very Dissatisfied	0	—
	Total	11	100%

The answers given to questions 52 and 53 by other respondents reveal a similar picture. Eighty-one per cent said that they were satisfied with their influence as chairmen or vice-chairmen (see Tables 8.4 and 8.5).

Table 8.4 *Other Chairmen or Vice-chairmen: Perception of Own Influence*

Influence:			
	Lot	7	47%
	Some	7	47%
	Little	0	—
	None	1	7%
	Total	15	100%

Not ascertained, 3.

Table 8.5 *Other Chairmen or Vice-chairmen: Satisfaction with Own Influence*

Attitude:			
	Very satisfied	7	44%
	Satisfied	6	37%
	Pro-con	2	12%
	Dissatisfied	1	6%
	Very Dissatisfied	0	—
	Total	16	100%

Not ascertained, 2.

From the above data a picture of the structure of power within the committees begins to emerge. Naturally, behind the quantitative findings lies a more complex reality which, in the remainder of Part Three, I attempt to explore.

The main committee meetings were a formality. The decisions were always those of the majority party. In most cases its view had been determined beforehand. On the few occasions when this was not so, majority party members would follow the lead of the chairman. Discussion in these meetings was kept to a minimum. The opposition members were allowed to go on the record and usually some attempt was made to reply to any criticisms they offered but no pretence was made of genuinely discussing issues. After committee decisions were taken they were implemented. Often this was the first occasion on which the public were involved.

How did the majority party come to agree upon the common line to be followed in committee meetings? For convenience I have divided this discussion of the committee decision-making process into two parts. First, in Chapter 9, I consider the stage of pre-caucus deliberation. The principal actors were the chairmen, some vice-chairmen and the chief officers. Other officers were also involved, but only at the discretion of chief officers. During this stage it was agreed which reports would go forward to the committee and in what form. The second stage was the pre-meeting (caucus meeting) of the majority party, which was normally held immediately before the main committee meeting. Majority party members went through the reports and considered the recommendations. Once a pre-meeting decision had been taken all majority party members had to support it in the main committee meeting whether or not they had been present at the pre-meeting. With only rare exceptions, officers were not involved in this stage.

NOTE: CHAPTER 8

1   The ten Labour councillors who were not asked this question either held no official position (five) or held positions to which little or no influence was attached (vice-chairmen of subcommittees with only very limited powers). (Not ascertained, one.)

Chapter 9

PRE-CAUCUS DELIBERATION

I like political control. It means I only have to sell an idea once;
then all we have to do is work out the details.
(a chief education officer in a Labour-controlled authority, quoted
in Jennings, 1977, p. 115)

The chief education officer should carry out committee decisions
cheerfully and efficiently whether or not his advice has been
taken ... 'Whenever possible his advice should be so given that his
employers do not realise they are getting it. They should be led to
the right conclusion under the impression that they are arriving at it
under their own steam.'
(Sir Arthur Binns, a chief education officer, quoted in Kogan and
van der Eyken, 1973, p. 48)

In discussing the pre-caucus deliberations I have been concerned with
two broad questions. First, I have been concerned with how the agenda
and the committee reports were controlled and secondly, with who controlled the record of decisions taken.

Committees operated mainly by considering reports presented by
chief officers. Control of whether or not particular reports should be
presented to committee, control of the timing of their presentation and
control of what policies they should recommend were the principal
functions of the committee chairmen (sometimes acting jointly with the
appropriate vice-chairman). I have therefore tried to establish how
chairman and vice-chairman functioned and specifically how they controlled (a) individual agenda items and (b) the content of the chief
officers' reports (including both the policy recommendations and the
information and argument offered in support of them). I was particularly concerned with point (b), and questions 47–51 in the questionnaire were designed to throw additional light on how chairmen and
vice-chairmen operated and how they saw their own role in this respect.

In the account which follows some respondents have drawn a distinction between two types of pre-caucus meeting: a pre-agenda meeting
and an agenda meeting. The former normally took place well in

advance of the main committee meeting and at these meetings chairmen and vice-chairmen were in a position to control not only which items would appear on the agenda but also the content of reports. Agenda meetings normally took place very near to the main committee meetings, usually on the same day just prior to the pre-meeting. The principal purpose of these meetings was for the chief officer to brief the chairman. But agenda meetings might also involve a discussion of the tactics to be adopted in the committee meeting (or even of the tactics to be employed by the chairman in the pre-meeting).

My second concern in writing about the pre-caucus stage was to ascertain who controlled the record of the decisions taken in committee and in particular how much control of the record was exercised by chairmen, since it was a job largely left to them by other Labour committee members. The questionnaire included questions about this (questions 44, 45 and 46).

According to Keith-Lucas and Richards, in the early years of this century senior officials in the large boroughs had considerable opportunities to exercise initiative. After the Second World War, however, their power was inhibited by the growth in the number of disciplined party groups, at least to the extent that some policy discussions took place at meetings they were unable to attend. However, this restriction on the power of officers has, they say, been 'masked and sometimes overtaken by a countervailing pressure – the growing volume of local authority business'. Increased scale and the growing complexity of business has meant that many authorities have had to delegate wider discretion to officers (Keith-Lucas and Richards, 1978, pp. 124–5). In the mid-1960s Harrison and Norton found that 'in nearly two-thirds of the local authorities' they consulted, 'officers were said to make a significant contribution to the initiation of policy and in nearly a quarter they were said to play the major part' (Maud Committee, 1967, Vol. 5, p. 196). At Wolverhampton Jones found that: 'Policy . . . was thought up by the officials, who then planted the policy inside the minds of the chairmen, who in their turn guided the committees' (1969, p. 177). What was the situation in Newcastle upon Tyne?

THE CHAIRMAN–CHIEF OFFICER RELATIONSHIP

In Newcastle the reports to committees could originate from many sources. They could be the result of complaints made by citizens or councillors; or of the identification of problems by officers or anyone else; or of the advocacy of some new proposal by a pressure group; or of promises made in party manifestos; or of previous committee decisions. For the outsider, the difficulty came in ensuring that a proposal was taken seriously, or that a complaint would be regarded as valid, or that powerful insiders could be made to share the outsider's definition of the situation. This was because the actual work of drawing up reports – which involved deciding what problems to recognise, defining these problems correctly and agreeing upon how to solve them – were activities which were carried out by civic centre staff under the control of their chief officers. The initiative for carrying out consultations lay in their hands. During the period of this study the extent of consultation increased but it remained inadequate. In particular, the committee reports themselves were not normally circulated in advance to citizens. They were sent to the press but they were embargoed until twelve noon on the day of the meeting, so that no member of the public could discover their contents before the committee meeting.

Not only were few outsiders involved in decision-making. Only a few councillors were able to play an active role; and even when a policy had been initiated by the party in power the effective deliberation was still dominated by officials. This point was made by one main committee chairman when, in 1974, he was asked by a district Labour party working group what had happened to the Labour party's manifesto policies about eight months after Labour had taken control of the authority:

We have fed them through. The problem is that evaluation of our policies is an entirely officer process . . . What happens when it's in the machine is that we lose influence over it. Reports often come out very different to the way they went in. There is very little political input during the administrative process. There should be more. It gets me that people think we control the machine when we actually have no input during the evaluation stage.

The most important decision taken by any committee was the annual budget. Yet chairmen often had to put in considerable effort to exercise a decisive influence over it. The same chairman also referred to problems which had occurred in preparing his committee's 1975/6 budget. He described it as 'a case of the Management Team laying down the law'. He continued:

They just sent a memo round the departments calling for estimates to be sent to the City Treasurer and for them to be based on zero growth. That didn't go to any councillor and was not discussed by Policy and Resources Committee. In fact it's never discussed esti-mates procedure. In most cases the estimates are prepared by officers. In [a named committee] we had some control over the esti-mates but some committees didn't have any. The committees don't discuss estimates till January. Well, by then it's too late to change priorities. Most chairmen don't know what's happening. Some chairmen don't know that the estimates have already gone to the City Treasurer. You wonder if you are really controlling what's going on.

A second main committee chairman had similar criticisms. He was interviewed immediately after the first chairman, quoted above, and was asked whether or not the claim that some chairmen were unaware that estimates had been submitted to the City Treasurer was true of his own committee. He replied:

Nothing has been done about estimates yet in committee. If I ask about estimates I get snippets of information. I did not in fact know that estimates had gone in to the City Treasurer from any department. I have not been told that my department's estimates have gone in. I'm not surprised that this has happened though.

At that time, Labour had not been in power long and Labour chairmen were still trying to assert themselves. Many were not very effective. But this was not true in every case. Two chairmen in particular did succeed in having considerable influence over the budget (the chairmen of the Education and Social Services Committees). Most chairmen increased the extent of their control over the departments as they gained experience in power. But problems remained and as late as the summer of 1977 the new leader felt that political control of the budgetary process was still inadequate. At a meeting of the district Labour party in July of that year, he said, while giving a report on the activities of the group, that he hoped backbench members of the group would be more fully involved in the future. He continued, 'In the past, some chairmen, let alone backbencher councillors, were not fully involved in the budgetary process'.

Although the initiative lay in the hands of the chief officers during the deliberation stage, they could by no means ignore the committee chairmen. One subcommittee vice-chairman put it this way: 'The chief officer always consults the chairman first. Otherwise he would be shot down.' A main committee vice-chairman put it slightly differently: 'The chairman–chief officer relationship is the dominant nexus', he said, but the chief officer 'needs to make sure that he and the chairman see eye to eye. They spend a lot of time doing that.'

Looked at from the chief officers' point of view, the chairmen were sounding boards. Each chief officer needed to obtain his chairman's consent to reports prepared by his department. A leading chief officer described his relationship very much in these terms: 'It is part of my role to be partly involved in politics, at least to the extent of holding informal consultations with the politicians before making suggestions.' One assistant director put it this way: 'The chairman is generally regarded as a sounding board. If the chairman opposes something then it's no good trying to put it through.' Or, as a second chief officer put it, 'There is no point in spitting into the wind'.

Chief officers, then, tended to prepare reports with the attitude of the relevant chairman in mind. What happened when a chief officer wanted to put forward a report to which the chairman was opposed? Much depended on personalities. Some chairmen took the view that their decision about what items appeared on the agenda was final; others allowed the chief officer a free hand. In still other cases an intermediate position was adopted. An example of the latter type was the relationship between the chairman and vice-chairman of the Housing Management Committee and the Director of Housing during the municipal year 1978/9. The chairman and vice-chairman exercised as much control as they wished, sometimes altering the content of draft reports, sometimes deferring reports, sometimes rejecting them outright. Usually the Director of Housing accepted their decision, but in two cases during 1978/9 he decided that he wanted to make recommendations to the committee to which the chairman and vice-chairman were opposed. The chairman and vice-chairman had an understanding with the director that where he felt strongly about something he would be allowed to put forward his recommendations in spite of their opposition. They allowed him to put forward the two reports in question and then secured the agreement of the Labour pre-meeting to reject them in the main committee meeting. A second, unspoken, part of the understanding was that the director should not make too much of a fuss in the main committee meeting when his report was being rejected.

Essentially, then, the chief officer needed the support and co-operation of the chairman. This meant that any chairman who allowed himself to be dominated was in large measure responsible for his own plight. (See Newton, 1976a, p. 66.) One councillor, the chairman of an important subcommittee who was also interviewed in 1974, referred, like the other councillors cited above, to the problem of officer domination, but he insisted that it could be overcome and that councillors who failed to overcome it only had themselves to blame. He was asked, 'Is there much resistance by officers to implementing party policy?', and replied:

The officers always say that they implement party policy. I asked [the then Chief Executive] if he had read the party policy document on housing modernisation. He had not read it. I told him to send copies to all the relevant officers and to tell them to read it. Some officers had to be put down with a fairly heavy boot at first ... We had to impress on them that policy was policy ... It's often said that there is a line between policy and detail. But the line should be where the politicians draw it ... If you don't control the detail in housing then you don't control the policy ... You've got to be fairly ubiquitous to ensure that policy gets carried out. When you want to speak to officers below chief officer they insist that you must deal with the chief officer because otherwise you would undermine his authority. [The Director of Housing] did that when I instructed an officer while he was at a conference in Scarborough. He got very upset. Sometimes they appeal to the Chief Executive or to the leader in an attempt to undermine the chairman.

Finally, he was asked, 'Why do councillors accept officer dictation?' and answered, 'Some councillors think it's their duty to support the chief officers. Too many see themselves as the dignified part of the apparatus. They settle for the deference of, say, the porters at the desk.'

Chairmen of this latter type were easy prey for officers determined to secure their agreement to some preferred outcome of their own. One reason why the ready agreement of such chairmen could be secured was that they failed to address themselves to the substantive issue in question. Dennis has described how the agreement of 'the indifferent influential' may be secured, as follows:

> The question to which the 'indifferent influential' addresses himself is not, 'What are the advantages and disadvantages *which will flow from the substantive decision* for me, my cause, my constituents, my supporters, my enemies?' It is, 'What are the advantages and disadvantages which will flow from my agreeing or disagreeing in the here-and-now with this officer, *quite apart from what I am being asked substantively to agree with?*'
> (1974, p. 167, his italics)

It is time to summarise. The principal participants in stage one, which involved defining problems and suggesting solutions, were the chairman and the chief officer. The initiative lay in the chief officer's hands and the chairman had to be vigilant to exercise control. But the chief officer could do little without the chairman's support. No committee decision could be taken without, at least, the chairman's acquiescence. In some cases such acquiescence was easily obtained but, essentially, the chairmen had the real power if they chose to use it.

In practice this meant that chairmen and chief officers had a symbiotic relationship. The chief officers were important because of their day-to-day running of their departments and because they controlled the effective starting point of every decision: they wrote the official reports. In particular this gave them the chance to control the information which all the key actors would use to take a decision. What factors explain the power of the chairmen in the relationship?

First, the official Labour Party view was that chairmen should play a very active role in controlling council departments. The Labour

Party's *Local Government Handbook* says this about the role of the chairman:

> From the party's point of view the role of the committee chairman is both executive and political. He or she is the key link between Labour group policy and administrative reality . . .
> He or she should work sufficiently closely with the officers to ensure that reports prepared for committees cover all points likely to be of concern to the group. Before politically significant reports are considered by the group . . . he or she should therefore read them in draft and if necessary redraft them so that they accurately reflect the party's intentions.
> (Labour Party, 1977b, p. 354)

Secondly, the power of chairmen in the relationship derives from the key functions which they carry out. The most comprehensive study of the role of chairmen is still that conducted by Harrison and Norton for the Maud Committee (Maud Committee, 1967, Vol. 5, ch. 7, pp. 136–90). They identified eight main roles which a chairman performs: (1) keeping discipline within the meetings; (2) acting as spokesman for the committee as a whole; (3) giving a lead to the committee; (4) making policy; (5) controlling the agenda and the reports; (6) taking decisions, for example, under 'chairman's action'; (7) acting as an advocate for the service provided by the committee; (8) co-ordination of the work of the committee with other committees, departments, or agencies. From the chief officer's point of view the most important of these was the chairman's capacity to give a lead to the committee, or less euphemistically, to control the committee. The majority party decided in private, without the involvement of officers, which way it would vote on the chief officer's proposals. In practice most committee Labour groups followed the lead of the chairman. This meant that when the chief officer was able to control the chairman then he could also control the majority party members of the committee and thereby the whole committee.

There is no legal basis for the great power of chairmen. Under Section 101 of the Local Government Act of 1972 a local authority can arrange for any of its functions to be discharged by a committee, a sub-committee or an officer of the authority. But it cannot delegate anything to a single member. It is common practice in many authorities for decisions to be taken by 'chairman's action' but these decisions require confirmation by the appropriate committee in every case.[1] However, in practice the absence of formal powers made very little difference in Newcastle. Indeed, for years committee chairmen have reigned supreme in many authorities. Lord Simon's study of Manchester City Council from 1913 to the mid-1920s reveals that, as he puts it, 'When a chairman has a forceful personality, is reasonably popular, and a member of the dominant party, he may become in time almost the Mussolini of his committee' (Simon, 1926, p. 95).

OFFICERS AND MEMBERS: THE WORKING RELATIONSHIP

The relationship between chairmen and chief officers was also affected by attitudes of Labour councillors towards officials. Herbert Morrison, while he was leader of the London County Council from 1934 to 1940, typified the attitude of Labour councillors who felt that officials should be kept at arm's length. He told officers and members 'not to become over-familiar and to fraternise with each other, so as to ensure that personal indebtedness would not influence public policy decisions' (Jones, 1973a, p. 5):

> Members were not to fraternise with the officials. They were told not to shake hands at County Hall, nor to address each other by their first names, nor to accept hospitality. All contact was to be at the LCC. (Donoughue and Jones, 1973, p. 196)

This view remains the official Labour Party line (Labour Party, 1976b, pp. 29–30). It is also reflected in the National Code of Local Government Conduct, adopted on the recommendation of the Redcliffe-Maud Committee (1974, pp. 46–9). Clause 5(ii) reads:

> Mutual respect between councillors and officers is essential to good local government. Close personal familiarity between individual councillor and officer can damage this relationship and prove embarrassing to other councillors and officers.

Consider now the attitudes of Labour councillors in Newcastle to officers. In the questionnaire councillors were asked about their attitude to local government officials. They were asked to say whether they agreed or disagreed with the following two statements:

> The councillor should keep his distance from the officers to make sure he will have no difficulty if he ever needs to criticise them or their decisions. (question 78)

> The councillor's job is to work closely in partnership with the officers in a close personal and friendly way. (question 79)

The following results were obtained. Thirty-five councillors answered question 78. Fourteen councillors (40 per cent) agreed. Of these fourteen, eight (23 per cent of the total) agreed strongly. Six councillors (17 per cent) had no strong feelings either way and fifteen councillors (43 per cent) disagreed. Of these fifteen, five (14 per cent of the total) disagreed strongly. The following councillors are typical of those who agreed with question 78:

Councillors would end up being subservient. (a backbencher)

I agreed [but] I can criticise the bastards whether I'm close to them or not. (a main committee chairman)

Councillors who disagreed with question 78 had this to say:

I'm a strong believer in civility. You don't get far if you demand. You should be close to them. (a backbencher)

There's no need to treat them like a leper. I have no difficulty in criticising them. (a vice-chairman)

Thirty-two councillors answered question 79. Seven (22 per cent) councillors agreed. Of these, four (12 per cent of the total) agreed strongly. Six councillors had no strong view and nineteen (59 per cent) disagreed. Of these, five (16 per cent of the total) disagreed strongly. Councillors who agreed with question 79 said:

You never get anywhere if you don't. If you get the officers' backs up they make it awkward for you – that's been my experience. (a backbencher)

Officers and councillors should work closely together till it goes against the political grain. (a backbench councillor)

Councillors who disagreed with question 79 said:

You've got to work with them but you should not get too close because your opinion can be coloured. (a subcommittee chairman)

I work closely with them so I know what's going on and that's all. (a chairman)

CONTROL OF THE AGENDA AND REPORTS

How many chairmen in Newcastle took advantage of their potential power to control the committee agenda and reports? Some quantification of the results is possible; indeed the questionnaire was designed to facilitate this; and the quantitative findings are set out below. They are followed by two examples, the first concerning a chairman in a relatively strong position and the second, a chairman in a weaker position.

The potential of each chairman to control the content of reports was clearly limited by whether or not he received draft copies of the documents before they had become the settled view of the chief officer in question. Chairmen were asked about this in questions 47 and 48. Five received genuine draft copies as a matter of routine, three only received them when they had been printed in sufficient quantity for the whole committee and one did not receive drafts at all. He merely discussed proposed policies with the chief officer. (One chairman did not answer this question and the chairman of Policy and Resources Committee is dealt with separately.)

A second consideration was whether or not there was any formal machinery designed to increase the chairmen's control. Specifically I have in mind the pre-agenda meetings because chairmen who held pre-agenda meetings increased their opportunities to control the content of chief officers' reports. Of the ten chairmen, seven held pre-agenda meetings and three did not.

A third factor was the extent of the facilities available to the chairman within the department. Some chairmen had no base at the civic centre and if they wished to see their chief officer would make an appointment with the official in his office; they had nowhere to work or hold discussions in private. Other chairmen had offices in the civic centre. This gave them a distinct advantage. Of the ten chairmen, in autumn 1977 six had rooms in the civic centre and four did not.

Example 1

Before the interview in 1977 this chairman had also been interviewed in November 1974. At the time, Labour had been in power for only eight months and the chairman's principal concern was to get control of the department and to change its direction after seven years of opposition rule. I have set out what the chairman said about this problem in 1974, before going on to consider what the position was in autumn 1977.

The chairman was asked, 'Where does power lie in local government?' and answered:

> I'm still looking for it. It's a misconception to think that when you're in power you run the council. There is a vast bureaucratic machine which goes on regardless. It's very hard to get it to change its direction. The committee is highly ineffective ... In my experience the main tactic [adopted by officers] is the positive policy recommendation, rather than refusing to implement something. They have the idea that they are pushing the authority forward rather than slowing down political instructions. They may play down things they don't regard as important ... They also play on the time factor. Most

councillors don't really have the time to go into things fully and the officers know that. I find myself moving in ever decreasing circles . . . I'm becoming more and more deskbound. Most of my consultation takes place with officers, behind the scenes. It's not entirely a war. I get more help than other chairmen, though in the past other chairmen have not demanded as much say. Another problem is time. At the last meeting an extra report was introduced an hour before the meeting and explained to me in the car. There is often not enough time to consider things properly. At present we're not the boss at all, the chief officer is.

These were the chairman's remarks in 1974. Consider now how the committee functioned about three years later. The chairman said that he discussed draft committee reports with the director 'very early on'. There was also a meeting between the director, the deputy director and the chairman and vice-chairman at which draft reports were discussed. The vice-chairman, however, felt somewhat excluded. He said that the chairman also met the director and deputy director without him present: the chairman 'assumes some powers. It's not a perfectly demo-cratic way of doing it, but then he's been chairman longer.' By autumn 1977, the chairman had become satisfied with the extent of his control: 'The director knows there is no point in putting up reports which would be thrown out. He tends not to bring reports we would be drastically opposed to.' When asked how often he changed the draft reports he said, 'The average report going to committee I modify rarely. On the run-of-the-mill reports, very rarely indeed. Something like the . . . budget I alter a great deal.'

By 1977 this chairman felt very much more satisfied with the extent of his control of the department than he had in 1974. But he still felt that there were problems. For example, he expressed frustration with the problem of getting adequate information. He put it this way: 'This problem of getting the right sort of information is one of the continuing frustrations of local government. Often facts are not in the form in which they can be used politically [that is, for policy-making].'

The chairman's position was considerably enhanced by the fact that he had been allocated a room in the department. The director had initially been very strongly opposed to this, arguing that it would confuse the chain of command and that it would make the chairman the executive head of department, in place of the director. But the chairman had insisted and was given his room, along with part-time secretarial help.

Although the chairman felt reasonably happy about his control of the department some committee members felt that there was a problem. For example, one main committee vice-chairman said:

We have to work twice as hard to change the officials as I expected.

They are entrenched, unless we put forward a very dramatic change and force it through. On other items they are entrenched. It's very easy for them to capture the chairman and the vice-chairman.

A very similar point of view was also held by a senior officer in the department. In a conversation in August 1974 he made the point that politicians only really had an effect when the party in power changed:

Political influence . . . is felt only in times of political change. In the old authority there was not much conflict till a year ago. The change of party meant changes. It had been fairly stable before that. When this present lot get established then that will change and things will settle down again . . . Basically in the old authority the officers tended to get their way much of the time. When things settle down it will be the same again.

Consider now how backbench committee members regarded the chairman–chief officer relationship. This is what some of them said when asked about the power of the director (questions 32, 34 and 36):

Most of the time you've just got to accept his advice. But he's not one hundred per cent effective. Decisions are made at the pre-meetings and some go against him. (a backbench councillor)

Occasionally [the director] has to put some effort in but usually none. He's a very strong-minded man. He's extremely articulate and knows what he wants. He's very difficult to trap or trip. The chairman and the director almost sleep together. The director is the dominant partner although the chairman would disagree with me. (a main committee vice-chairman)

One committee member regarded the director as the real power behind the scenes. The director, he said:

. . . is unseen but one knows that the chairman is expressing the chief officer's views. He says it with great confidence but you know it's not him that's talking. It's the shadow behind. He is the most powerful chief officer I would say. His influence during the committee meeting itself is not marked. It lies behind the meeting.

Example 2
This chairman did not receive draft reports. Instead he held an agenda meeting with the chief officer about a week before the main meeting. He described what happened at these meetings as follows: 'The chief officer would normally talk about a problem with me. He would

explain it and put it in the terms that there was no alternative. He would seek my support. Once it had been given my approval he could bring it to committee.' The vice-chairman was not consistently involved in these meetings.

One leading committee member was very critical of the chairman: 'The chairman has no effect at the present time. It's the officers that are carrying out policy. The chief officer has too much effect. He should get a lead from the chairman. The chairman should really go into it thoroughly and give the officers a lead.' There was a good deal of truth in this criticism. The chairman did consistently fail to give the officers a clear lead. This was because he felt that it was not his job to do so. His attitude is revealed by the following example. At a pre-meeting of Labour members of the committee held in September 1977, there was a discussion about the delay in receiving some final accounts. The delay was being blamed on the Chief Executive. The chairman proposed that Labour members should press the officers to get on with preparing the final accounts in the main committee meeting. A senior committee member then suggested that the chairman should chase the officers up between the committee meetings. The chairman replied, 'It's not for voluntary workers to go telling officers to do a thing like this. I wouldn't tolerate it in my job.' (He was a skilled manual worker.) The committee member protested, 'But it's the chairman's job. You've got to be on top of your officers and tell them to do things.' The chairman, however, refused to accept this.

CONTROL OF THE RECORD

To what extent did chairmen control the record of decisions taken by their committees? They were asked three questions about their control of the record (questions 44, 45 and 46). Only two chairmen said they received draft minutes. Furthermore, they both said that they modified these draft minutes only rarely. This meant that this function tended to be controlled by officers, although the more powerful chairmen were in a position to insist on changes when they were determined to do so. The extent of the chairmen's control of the minutes was a somewhat controversial issue within the authority. Haydn Richards in his account of a conference of local government officers reveals some of the reasons for this:

At a recent...conference of officers...we listened to a learned paper about council minutes...Suddenly, halfway through the paper the audience gasped as if hit by a torpedo amidships and in muted shock they laughed, some sympathising but many criticising. Here it was again...! The speaker was saying how it was his council's practice to send draft minutes to the committee chairmen

and the council leader for vetting.
(Richards, 1977, p. 1287)

Clearly, for some officials this degree of member control is laughable. Many officials in Newcastle had a similar attitude. One main committee chairman mentioned that he had been in dispute about access to draft minutes: 'I had a big argument with the Director of Administration on this', he said. 'I wanted to see all the draft minutes. I've done it on occasion but mostly I don't . . . I get the minutes with the agenda and papers.' Until the time of this chairman's interview the Director of Administration had opposed any chairman who tried to obtain draft copies of minutes. Draft copies were circulated to officials in all departments involved in meetings to enable them to comment yet the Director of Administration refused to circulate them to committee chairmen on a regular basis. At about this time, however, more and more chairmen were demanding access to the minutes and the Director of Administration was forced to relent. A second chairman said during interview that he did receive draft copies of the minutes but added: 'I had a hell of a battle to get it . . . The Director of Administration was the difficulty. He argued it was not the chairman's responsibility to do it, that they could be changed at the meeting. It took two or three months to get it.'

At the time of writing any chairman who wanted to could obtain draft copies of minutes.

NOTE: CHAPTER 9

1 Power could be delegated to as few as two members, if they formed a subcommittee. In Newcastle it was quite common for such subcommittees to be established and for them to be made up of the chairman and the vice-chairman.

Chapter 10

THE LABOUR PRE-MEETINGS

In preparing their recommendations to the politicians, planners normally take a fairly lengthy run-up. There is research to be done and reports to be drafted and then redrafted; working groups are set up to study particular issues and identify the possible options. Professional opinions are traded back and forth, possible recommendations are passed up the hierarchy and vetted . . . At the end of the procedure there emerges the departmental view around which all will rally. . . .

The local politicians rarely get the chance for such mature consideration. The appearance of the planners' recommendations on the committee agenda is all too often the first point at which the politician begins to consider the issues involved . . . the party groups may only meet . . . for half an hour or an hour before the committee meeting . . . Given the short and infrequent nature of group meetings in many authorities . . . decisions cannot always be considered in the detail that might be desirable.
(John Gyford, describing his experience as a member of the Essex County Council Planning Committee (1973–1977), 1978, pp. 5–6)

Our criticism was . . . directed . . . at a system that appeared to hold back our efforts by involving backbench members in mounds of paper and extensive committee work, yet removed them from real decision-making.

These were the words of Councillor Bryn Davies, deputy leader of Lambeth London Borough Council. He reports strong dissatisfaction among Labour councillors in Lambeth with the failure of the council machine to provide a more effective say in policy-making for elected representatives (Davies, B., 1979, p. 11). What was the position of the backbench councillor in Newcastle upon Tyne?

In Newcastle the backbench councillor's main chance to exercise influence on committee decisions came during the pre-meetings. Before considering the pre-meetings in more general terms I have set out below two examples of the way in which pre-meetings functioned.

Initially this committee's pre-meetings were held for only three-quarters of an hour immediately before the main committee meetings but this was extended to one hour during 1977. A principal feature of the caucus meetings was that they were dominated by the chairman. Some committee members made this point during interview: 'It's not common for the chairman to be defeated' in the pre-meetings, said one main committee chairman, 'I've not known a case where it's happened.'

A second informant (a subcommittee vice-chairman) was also extremely critical about the power of the chairman and offered the following explanation:

> Against the chairman the backbencher has got no chance . . . The chairman and the vice-chairman have got all the gen. This chairman, if you challenge him in the pre-meeting or the main committee he gets highly annoyed and aggressive. It doesn't lend itself to solving problems . . . You can't challenge the chairman if a report is just presented at committee . . . The chairman may call you an anarchist or a communist to try to belittle you . . . No one knows the powers of the chairman. It's not clear in standing orders what they can or can't do.

He also referred to two further problems. First, he said that 'the chairman should not accept a long agenda'. The officers, he asserted, 'use this to steamroller reports through'. Secondly, he said that the chairman's position was strengthened by the role which the vice-chairman played. He felt that the chairman and vice-chairman were 'two of a kind: the vice-chairman backs the chairman all the way. He follows the chairman so that in a pre-meeting you are fighting two of them.'

Clearly these committee members felt that the pre-meetings made little or no difference to the policies of this committee. How true was this? I measured the effectiveness of the caucus meetings over a twelve-month period (from July 1976 to June 1977) by examining how many of the resolutions adopted by the committee had been successfully recommended by chief officers (more than one chief officer presented reports to this committee) and supported by the chairman and vice-chairman. The following results were obtained.

The Effectiveness of the Pre-meetings: Eleven meetings were held and at these meetings officers put forward 424 recommendations. Of these, 420 (99 per cent) were accepted and four (1 per cent) were either not accepted or amended. In addition to this, forty-six other resolutions not specifically recommended by officers were adopted by the committee.

Thus, in total, 470 resolutions were adopted by the committee. Of these, 420 (89 per cent) had been successfully recommended by officers and supported by the chairman. Only fifty (11 per cent) were a consequence of the activities of the councillors on the committee. (The majority of these were simple additional resolutions which were not opposed by the chairman and the chief officer.)

The key factor was the support of the chairman. Unless a backbencher could secure the chairman's support he had, in practice, almost no chance of being effective. No decision of this committee was taken during the two-year period January 1976 to January 1978 without the chairman's consent.

Possibly the most disturbing aspect of the way in which the caucus meetings operated was that many issues were simply not discussed in anything approaching a reasonable, let alone thorough, way. Attempts to discuss a report seriously were often met by impatience and hostility. Few councillors wanted to go into issues thoroughly. If anyone else wanted to, then their inclination was to stop them. The attitude adopted towards the reports of a subcommittee illustrate this.

The subcommittee was chaired by a member of the main committee, a backbencher of considerable experience. When the backbencher was made chairman in May 1976 it began to work more effectively than in the past. They went into issues thoroughly and sometimes came to conclusions different from those of the department. This was not only resented by the department but also by the chairman of the main committee. For example, at the November 1977 meeting of the committee the chairman said, while referring to the minutes of one of the subcommittee's meetings, 'When will we be rid of this scourge?'

Several backbench committee members shared his general attitude and contributed little to the work of the committee. One committee member, a main committee vice-chairman, referred to the weakness of some backbenchers as follows: 'Few backbench councillors are capable of carrying out the executive function . . . A lot of councillors don't even read their papers. That's how the officers can influence things. If you don't see things for yourself you can be influenced.'

The impoverished nature of the discussion of reports by pre-meetings was not only due to the attitude of participants. The time allowed for the pre-meetings was inadequate. This made it very difficult to bring any new information to bear on an issue. This is how one subcommittee vice-chairman described the problem of introducing new information in a pre-meeting:

You would be at a loss because if you come in like that you always seem to be stirring it up . . . You may have some extra information which they haven't got but some colleagues would try to make you look like a stirrer and this would be accepted by the others. You could be right [on the substantive issue] because you have more information and they could be wrong. But it would make no difference . . . There should be time to show the information you've got is authentic . . . Because of the time factor the chairman says 'We haven't got all day, we must get through the agenda'.

The following is an illustration of how thoroughly issues were discussed in committee pre-meetings. I timed four meetings that took place during the period from September to December 1977. They are representative of the other meetings and if anything cast a favourable light on the way the caucus meetings were organised.

A total of 156 items were on the agendas. Seven were not discussed (all at the September meeting) because of a lack of time. The remaining 149 were discussed in 167 minutes at an average of one minute seven seconds per item. The greatest length of time devoted to any item was twelve minutes and the second greatest was eight minutes. The shortest was a second or two: literally the time it took for the chairman to read out the heading of the report and for members to chant back 'Agreed'. Often a meeting would speed through a string of agenda items in the time it took the chairman to read out successively the heading of each report, put the ceremonial request, 'Agree the report?' and for the backbenchers to chant back, 'Agreed'.

The twelve-minute and eight-minute discussions both occurred during the September pre-meeting and it was because this length of time was devoted to these items that seven reports were not discussed at all. This pressure partly explains why members behaved in an impatient way in pre-meetings. They realised that if anyone pressed an issue, and discussion exceeded three or four minutes, then the meeting would not get through the agenda in time.

An obvious rejoinder to the criticism that inadequate time was devoted to discussing reports is that many of the reports simply did not merit it. There is some truth in this. Many reports were routine and completely uncontroversial; some followed automatically from previous decisions; others had already been agreed 'in principle' to the extent that they formed part of a programme which had been approved earlier. However, the degree of discussion devoted to reports bore no relationship to the importance of the agenda item. Huge expenditures might be agreed as readily as minute expenditures. Sometimes a heated discussion took place on an item about which nothing could be done, or which had already been agreed in principle. Thus, the objection that some reports did not merit discussion was valid, but it could not be said that time was only devoted to the consideration of reports which required discussion; indeed often the reverse was true. This is illustrated by the way in which the committee dealt with the most important decision of the year: the annual budget. I discuss below how the budget was agreed for the financial years 1977/8 and 1978/9.

The 1977/8 Budget

The involvement of backbench Labour members of the committee in preparing the budget for this year was negligible. The only opportunity they had to contribute occurred during the pre-meeting which was held for three-quarters of an hour immediately before the main committee's special budget meeting. It took the form of officials explaining the budget to Labour members. Not a single change was made to the document. The main meeting was held immediately afterwards and the chairman rushed through the business in record time. The minutes record that it took fifteen minutes but actually it was nearer ten. The chairman was proud of this achievement and subsequently boasted about it on more than one occasion.

The 1978/9 Budget

The procedure for considering the 1978/9 budget was a considerable improvement over that for the previous year, largely as a result of the new leader's attempts to improve the involvement of backbenchers in decision-making. However, it was still seriously defective, as I show below.

In November 1977 a special evening meeting of Labour members of the committee was held to discuss a new three-year programme. The three-year programme began in the financial year 1978/9. This special caucus meeting was the only occasion on which policy was discussed by the Labour members of this committee without a time limit. As a result the three-year programme was discussed in a fairly leisurely way and a small number of modifications to the draft were made. I shall refer here to only one example, namely, that provision in the 1978/9 capital programme for one item was increased from seven to ten.

The fact that this meeting took place was a considerable improvement over the 1977/8 procedure. Nor was it the only improvement. The 1978/9 proposals were discussed at a further caucus meeting. The document sent out to Labour members for this second meeting was what they had agreed as year one of the three-year programme. The caucus meeting called to approve the 1977/8 budget had been held immediately before the special budget meeting of the committee and this made modification virtually impossible. By contrast, the 1978/9 proposals were considered more than a month in advance of the special budget meeting: at the regular December pre-meeting. The document had only been distributed to Labour members and this made amendment more easy because Labour members felt less constrained (rightly or wrongly) when acting in private.

However, the way in which this pre-meeting discussion was arranged cancelled out this potential advantage. An error was made about the time. The normal pre-meeting was due to start at 1 p.m. and the full committee meeting at 2 p.m. The letter, sent out with the budget papers, said that the budget would be discussed at 1.45 p.m. It should have said 12.45 p.m. which would have allowed an additional fifteen minutes to discuss it (itself inadequate) but because of the error the regular pre-meeting (also already too short) had to be shortened. The budget was discussed from 1.38 p.m. to 1.51 p.m. and again took the form of an explanation by officers. However, no Labour councillor objected since there had been the meeting on the three-year programme and the budget document before them was what they had agreed on that occasion for 1978/9.

Events then took a new turn. Between the regular December pre-meeting and the special budget meeting in January 1978 the chairman agreed to modifications in the budget document adopted in December. He did so without telling committee members. Indeed when members received the final version of the budget (in January) attention was not drawn to any modifications. One change which the chairman had agreed to was that no finance at all was included in the capital pro-gramme for the provision of the item referred to above.

Thus, to summarise, the procedures had been improved and Labour members had made limited use of this to modify the budget for 1978/9. But the improved effectiveness which they had gained had then been cancelled out by the chairman's unilateral action.

CASE STUDY 2

Pre-meetings of the Direct Labour Organisation Committee were not always held. When they were, they were held for an hour or half an hour before the main committee meeting. They were not very effective. On the occasions when pre-meetings were held, the chairman often raised issues not contained in the official reports. No notice was given of his intention to do so, and usually no additional documents were circu-lated. Even issues of the greatest significance were considered without notice being given. For example, in January 1978 a decision was taken which fundamentally altered the structure of the direct labour building organisation.

Labour members were asked to remain behind after the regular meet-ing and a three-page document was then put round the table. Before anyone was able to look at it, news came that a previously scheduled joint meeting of the Housing Committee and the Housing Management Committee, called to discuss the annual budget, had just begun and that the opposition had a majority. Several Direct Labour Organisation Committee members were involved and had to leave for that meeting. They returned after about half an hour to find that the document effectively recommended winding up the DLBO as it was then constituted. An attempt was made to initiate a discussion of the problems referred to in the report and to make one or two positive suggestions but few of the people present were willing to allow this. A backbench member of the committee argued that the report was not positive enough and suggested that it might be an idea to try to combine the – then topical – debate about industrial democracy within the authority, with the discussions about the future of the DLBO due to take place with the unions. The chairman replied, 'We must let the managers manage. It's not our job to do this sort of thing. It's not the councillor's job to manage. I don't feel able to tell them how to do their job. They're more intelligent than me.' A main committee chairman interjected, 'Not more intelligent, just better educated', but it made no difference. It is often argued that councillors should not interfere in matters of detail but this chairman clearly thought that councillors should not interfere in matters of strategy either, for a decision about whether or not to raise the issue of industrial democracy within the authority was clearly strategic in character.

After only half an hour's discussion (the meeting lasted approximately one hour altogether) the report was agreed without amendment. This decision was taken without anyone being able either to read or study the document properly. Indeed, the chairman subsequently claimed that he had not seen it in advance either.

The measures taken at that time did not solve the problems of the organisation. At the time of writing the whole future of the organisation was very much in doubt because of the huge losses it had incurred.

CONCLUSIONS

The above examples reveal that some pre-meetings were generally not very effective. How typical were these examples? The committees operated largely by receiving reports from the officers. These reports normally concluded by recommending particular courses of action to the committee. Labour councillors were asked to state how often in their experience these recommendations were changed during the caucus meetings. The question was put as follows:

Still talking about the ... Committee. May we talk a bit about pre-meetings? How often in your experience are officers' recommendations changed during pre-meetings? Would you say *very often* or would you say *sometimes* or would you say *rarely* or *never*? (question 43)

The degree to which their effectiveness was curtailed was revealed by the answers councillors gave. Of the thirty-four Labour councillors who answered this question, twenty-four (71 per cent) said that recommendations were changed rarely. Ten (29 per cent) councillors said they were changed sometimes. None at all said they were changed often.

The inadequacy of the pre-meetings can be explained by four factors. First, the degree and form of access to information enjoyed by Labour committee members; secondly, the time at which the pre-meetings were held and the length of time allowed for them; thirdly, the calibre of individual councillors; and fourthly, the strong tradition of party discipline within the Labour party.

(1) *Access to Information and the Timing of its Release*

This is how one main committee chairman described the difficulty for the backbencher of influencing caucus meetings. He said, in answer to question 37, that the backbencher was 'giving himself an impossible task'. 'I would say', he continued, that the backbencher 'has *no effect at all* . . . His difficulty is getting information that will allow him to prepare his case, because it's confined to the chairman and vice-chairman and the bloody officers.' (My italics.) A second Labour councillor put it this way:

> The chairman and the chief officer become nearly one. One protects the other. The chief officer is able to cosset the chairman and give him information which is kept from the other councillors. The chairman then shields the chief officer from the other councillors . . . Chairmen are able to use information as a lever to get other Labour members to support them.

How valid was their claim that information was such an important factor? Certainly other studies have found that control of the flow of information could assume great importance. For example, Friend and Jessop observed that 'In Coventry, we were able to see much evidence of the powerful position which a chief officer could achieve through his control over the flow of information to committee' (Friend and Jessop, 1969, p. 55; see also Newton, K., 1976b, pp. 162–3). What was the position in Newcastle?

Most of the information which committees used to take decisions was contained in the reports presented by chief officers. These reports normally recommended a particular course of action and contained whatever information and argument in favour of the recommended proposal the chief officer, with the consent of the chairman, found appropriate. An important question was, how far in advance of committee meetings were these reports circulated? This affected (a) how long councillors had to read them, (b) how long councillors had to obtain explanation of the reports from relevant officers and (c) how long councillors had to gather extra information in order to test statements made or conclusions drawn in reports. Clearly, the shorter the period between the release of information and the meeting, the weaker was the position of the backbencher. Furthermore, was there any evidence that chairmen or chief officers deliberately controlled the release of information so that these constraints applied?

Some chairmen and chief officers did tend to control the release of information. They tended to release only information which demonstrated the rightness of the policy they supported. This was no accident. Attempts to obtain additional information were often vigorously resisted. Similarly, attempts by ward councillors to discuss draft reports were also resisted by some chairmen. I have set out examples below.

Several Labour councillors referred to the shortage of information which they suffered from and some suggested that it was deliberate: that some chairmen, or the chief officers, or both, controlled information to increase their own influence on committee decisions. As a matter of routine, committee reports were, as we have seen, released near to the committee meetings. The pre-distribution deliberation was confined, also as a matter of routine, to the director, some other officers at his discretion, the chairman and to a lesser extent the vice-chairman. However, if a councillor particularly wished to have access to a report before it was finalised would he have been prevented from doing so? What follows is an account of what happened when a ward councillor tried to obtain access to a draft copy of a report about an issue affecting his ward.

A backbench councillor asked to be consulted by officials on the preparation of a report which they were preparing at his request on the heating system in a housing estate in his ward. He had called a meeting in April 1976 to hear people's grievances and council officials had turned up and promised to prepare a report on the very inefficient and extremely expensive heating system. The councillor asked to be consulted during its preparation and this was agreed by the assistant director actually responsible for preparing the report. The councillor had suspected that the report would list various options and conclude that they were all too expensive – it did – and that this would result in no action being taken. He recognised that finance was in short supply and had wanted to see whether or not all possible alternatives had been considered, particularly cheaper ones.

In early July the councillor reminded the assistant director about the agreement to consult him, but nothing happened. At the end of the summer recess, on 31 August, he asked again. He was told that he could not be consulted without the permission of the chief officer. The chief officer was then approached. He said that he would have to ask the chairman. In the end there was no consultation and the report was duly put through the committee process on 8 September, when its recommendations were agreed.

The councillor raised the issue at the pre-meeting of the Housing Management Committee held on 8 September. The councillor told the meeting that he had wanted to see a draft copy of the above report so that he could make some suggestions about its contents and he reported that he had been prevented from doing so. He raised the general issue of ward councillors being consulted whilst officers were preparing reports on matters affecting their wards.

The chairman replied as follows: 'We ask the officers for a report and we get our chance to change it when it comes to committee.' He was supported by the vice-chairman of the Housing Committee. She said, 'If councillors were consulted before a report was brought [to committee] it would come as a report from the officers *and* the local councillors' and she concluded: 'there would be nothing left for the other councillors to do.' The vice-chairman of Housing Management Committee supported her: 'The problem is that not all the ward councillors are on the Housing Management Committee and so if the local councillors had been consulted in the preparation of a document they [the committee members] might be duped into thinking that there was nothing left to do.'

The chairman then added, 'We must also take into account the time element. It would inevitably take a lot longer.' The backbench councillor tried to press the issue and to reply to his opponents' arguments but the other members soon became impatient and time was pressing. The issue was dropped with an assurance that it would be discussed again. (It was not discussed again.)

The councillor pursued the matter by asking the chief officer for an explanation of his failure to consult him. The Director of Housing gave the following explanation:

> When you first asked to be given a copy of the report before the committee meeting, [the assistant director] referred the matter to me. I then discussed your request with the chairman who felt that any early distribution of a report would be contrary to the spirit of the procedure which already exists, i.e. that all committee members should have the opportunity to discuss any report submitted for their consideration on an equal footing. He therefore recommended that I should withhold the report until the distribution to committee members.

After a couple of months the councillor wrote to the director again. He raised the general issue of consulting ward councillors:

> Dear Sir,
>   Thank you for your letter . . . concerning the consultation of councillors before committee meetings. I would be grateful if you would provide me with a little further information:

(1) When a report is being prepared on a matter affecting a particular ward, what arrangements are there for the councillors representing that ward to comment on the document before it is finalised?

(2) If the answer is that there are no arrangements, then do you have plans to introduce any?

Yours faithfully,
(signed)

The Director of Housing replied as follows:

My answers to the questions raised in your letter . . . are:

(1) There are no arrangements that I am aware of whereby an officer seeks comments from ward councillors when preparing reports concerning a particular ward. . . .

(2) I do not have any plans to introduce a system of consulting ward councillors about reports. I am responsible for reports I prepare for committee and I see a distinction between a ward councillor having an opportunity to make his views felt to committee and in some way commenting on a draft report other than on a question of fact. . . .

Yours sincerely,
(signed)

Thus the ward councillor's attempts to be consulted were strongly resisted. The chairman and the director had succeeded in restricting the ward councillor's access to information (that is, to the committee reports containing the facts and the director's recommendations) until a point at which it became very difficult for him to have any effect on either the facts or the conclusions drawn from them.

A second example involved the chairman of the Housing Committee. The performance review subcommittee had set up a housing working group which was carrying out work on the housing action areas. It had asked for a copy of the annual report on progress in the housing action areas which the Housing Department had to present to the Department of the Environment. This gave the position up to the end of June 1977. It had been requested by the working group in July but by the end of September it had not been received. An assistant director, it turned out, had refused to release it on the ground that it had not been presented to the Housing Committee. He had not communicated this reasoning to anyone. He had just ignored the request for the document. When challenged, he argued that the intention was to present the report to Housing Committee and that it could not, therefore, be seen by any councillor before it was distributed to Housing Committee members. Thus the document was widely circulated amongst officers at all levels within the Housing Department and had been sent to the Department of the Environment as the council's official progress report and yet it was not available to individual councillors, nor to an official working group of the council. The officer was asked whether it could be made available to councillors if it was *not* going to be presented to the Housing Committee. He replied, 'Yes, I suppose it could'. Nevertheless, he claimed that it would require the chairman's consent to release it in advance of the normal distribution date (which, at that time, was not known). As a result, the working group chairman asked the Housing Committee chairman to release the document. She replied, 'Well what are you going to do with it?'

'Read it', the councillor replied.

'Yes, but what are you going to do with it?' the chairman insisted.

'Well, that depends on what it says. Maybe nothing. It depends on what the document says.' The councillor explained that he just wanted to read the document to find out what the position was, but the chairman appeared to be incapable of understanding this.

She replied, 'Well, I want to see it first. When I have seen it I will decide whether or not you can have it.'

In fact after seeing the document she agreed to release it. But by this time it had taken two months to get it. This example is all the more disturbing because of the nature of the information. It was basic factual information which the Department of the Environment required by statute. It demonstrates that once secrecy becomes conventional it extends easily to information of the most routine kind.

The problem for the councillor was not only one of restrictions on access to information; when access was permitted the timing of its release was also important.

The length of time that reports were distributed in advance of committee meetings varied. The longest period was six and a half days. Usually it was less. This created serious difficulties.

First was the problem of simply reading the reports. As one main committee chairman put it: 'You get the documents too near the meeting so you're not allowed to read them. You're not allowed to understand them.' A main committee vice-chairman said that he had special difficulties because he worked shifts. Papers for his committee were distributed on the Friday for the meeting on the following Tuesday. His only real opportunity to read the reports came on the Sunday. If for some reason he was unable to read them on the Sunday he often had no opportunity to read them at all.

Several other Labour councillors mentioned the problem but felt that it was a matter of personal default. They strongly condemned fellow Labour councillors who failed to read the reports. A main committee vice-chairman, when talking about the influence of officers, said, 'A lot of councillors don't even read their papers. That's how the officers can influence things', and a backbench councillor said, 'At least I do read the papers. Unlike some who don't and then just throw them away.' Jones found a similar problem in his study of Wolverhampton: 'The Labour councillor was supposed to examine the agenda and thus come primed to the . . . group meeting, but complaints were frequent that councillors opened their agenda envelopes only on arrival at the group meeting' (1969, p. 174). This problem gave rise to a standing joke (which was not without foundation in fact) about one main committee chairman who, as a backbench member of other committees, was said invariably to preface her (frequent) remarks with the statement, 'I haven't read the report *but* . . .'.

Secondly, it only left a short period during which councillors could have unclear parts of the documents explained by the officers who had written them or could check particular facts with officials. It might be thought that explanation could be obtained at either the pre-meeting or the main committee meeting and that therefore the shortage of time in which to discuss reports with officials beforehand was not a major difficulty. This might have been so if both pre-meetings and the main committee meetings had not operated as they did. Certainly, some explanation could be obtained in both pre-meetings and committee meetings but it was, in practice, nearly impossible for a councillor to employ any information or understanding so gained in order to challenge the recommendations in the reports. It was not possible in committee meetings because Labour councillors were not allowed to oppose the line agreed by the pre-meetings. It was not possible in the pre-meetings, because when a chairman wished to prevent it, he could normally do so.

Thirdly, councillors also had only limited time in which to gather additional information – perhaps to challenge or at least test a statement in a report. Some additional information could have been gathered by asking questions in either the pre-meetings or in the main committee meetings. In the case of the main committee meetings, this option was of limited value. This was because hostile and persistent questioning by Labour members in the main meetings was strongly frowned upon and indeed discouraged not only by chairmen but by most Labour councillors. Hostile or critical remarks were supposed to be confined to the pre-meetings and attempts to the contrary were strongly resented and resisted. There were one or two Labour councillors who occasionally did this and they earned considerable hostility as a result. One backbench councillor, speaking of a particular committee, criticised this practice during interview. 'The problem', he said, 'is that some Labour members go to the committee and make a fuss about something without going to the pre-meeting first. They embarrass the party and ask awkward questions.' Thus Labour members were supposed to confine hostile questioning or criticism to the pre-meetings but, since officers did not attend these meetings, backbenchers were heavily reliant on what the chairman was able or willing to say. Not all chairmen were able to deal with the comments and criticisms of their committee members but the majority were.

These problems put the backbenchers in a weak position and meant that reports favoured by the chairman were likely to go through unchallenged unless something quite out of the ordinary happened. Such difficulties are not confined to local government. Joe Haines reports that the Chancellor only told the Cabinet about his budget some twenty-eight hours before he presented it to Parliament. 'On that time scale', he says, 'only a massive spontaneous revolt against the Budget can make a fundamental change in it' (1977, p. 66).

This was why the timing of the release of information was important in Newcastle: for a report to be changed, something like Haines's massive spontaneous revolt was necessary. This point was made by more than one respondent during interview. In the pre-meetings, according to one backbench councillor, 'the backbencher has little effect except when there is an explosion'. Or, as a main committee vice-chairman put it, 'Unless the pre-meeting is bloody about something' it usually agrees to support the chairman.

Councillors were specifically asked about gathering information in addition to that contained in the officers' reports in question 29. These answers have been aggregated and the following results were obtained. Each councillor answered the question for between one and five committees. The total number of responses was 121. Of these, seventy-nine (65 per cent) were positive and forty-two (35 per cent) were negative. Councillors who answered question 29 positively were then asked question 30. The results are set out in Table 10.1.

Table 10.1 *The Extent to which Labour Councillors Gathered Information in Addition to that Presented in Official Reports*

Lot	20	26%
Some	28	36%
Little	28	36%
None*	2	3%
Total	78	100%

Not ascertained, 1.

*Two councillors responded by saying that they found it necessary to gather additional information but that they were unable in practice to do so.

Thus, seventy-two of the answers were either a straight 'No' or, if initially 'Yes', that only a little or no time was actually spent gathering extra information: that is, 60 per cent. The following are typical comments made by Labour councillors while responding to this question:

> I find it necessary but in practice I don't spend any time on it in most cases. You are not given enough time to digest reports. So you go into committees at half-cock. (a backbench councillor)

> I would if I had the time. It's difficult to know whether or not the reports are a truthful account of the matter. I am unable to spend more than a little time. (a main committee vice-chairman)

This problem made many Labour councillors very reliant on the chief officers' reports for their information. This is confirmed by the answers to question 64. Councillors were asked to state what their main source of information was. The results are set out in Table 10.2.

Table 10.2 *Labour Councillors: Their Main Source of Information*

Main source of information	Number	%
Chief officers	13	41
Constituents	10	31
No main source	6	19
The group leadership	1	3
The local party organisation	1	3
Other	1	3
Total	32	100

Not ascertained, 8.

Thus 41 per cent of Labour councillors in Newcastle said that chief officers were their main source of information. Dearlove found a very similar result in Kensington and Chelsea. Forty-four per cent of councillors in that authority said that officers were the main source of information which they used for decision-making. Indeed, a further 10 per cent of his respondents said that agendas were their main source of information (1973, p. 179). As Dearlove points out, officials 'in many ways serve as links to the "real" world, filtering in what they feel their political masters need to . . . know' (p. 176).

Many of the Newcastle Labour councillors, who were highly dependent on chief officers' reports, were also very critical of them. Two sets of questions were asked about chief officers' reports.

Councillors were asked whether or not they found officers' reports too technical (question 72) and, if so, how often this happened (question

73); and secondly, whether or not they felt that officers failed to present alternative policies when they should have done so (question 74), and, if so, how often this happened (question 75). Answers to questions 72 and 74 were overwhelmingly positive.

All forty respondents answered question 72. Thirty-five agreed with the statement and five disagreed. Of the thirty-five that agreed, twenty-nine (83 per cent) thought this happened often or sometimes. Only six (17 per cent) thought it happened rarely. One main committee vice-chairman who agreed with the statement said: 'I agree. Finance is a waste of time. I can't make head nor tail of it. It may be because I'm thick. I could be a dunderhead. I don't know.' A number of councillors had a similar attitude, particularly towards financial matters. That many councillors had difficulty in understanding the reports clearly made their dependence on them for information all the more serious.

Thirty-eight Labour councillors answered question 74. Thirty-seven (97 per cent) agreed and only one councillor disagreed. Of those that agreed thirty-five answered question 75. Of these thirty-five, sixteen (46 per cent) said it happened often, fifteen (43 per cent) said it happened sometimes and only four (11 per cent) said it happened rarely: that is, 89 per cent thought it happened often or sometimes. The following two councillors are fairly typical of most councillors who agreed with the statement.

Yes, they often only put down their own opinions. It should be their duty to put down alternatives. (a subcommittee chairman)

This is a serious problem. One of my complaints is that we are *never* presented with alternatives. (a main committee chairman)

This problem tended further to strengthen the hold of the main committee chairmen and of the chief officers.

(2) *The Length of Time Allowed for Pre-Meetings*
The pre-meetings were held immediately before the main committee meetings. This led to two problems. First, it indirectly meant that discussion was inhibited; secondly, it placed a time limit on the meetings.

The pre-meetings met when the reports had already been distributed to the opposition and the press. It was expected by the opposition and the press that the reports would be supported by the majority party and therefore it was always felt by Labour councillors that it would be embarrassing to withdraw or amend a report. A good deal was made of this problem, particularly by the chairmen. However, their real worry was not so much that the party would look foolish in the eyes of the press and the opposition but that they would look foolish in the eyes of their chief officers. A chairman was expected by most chief officers to be able to secure support from his pre-meeting. Indeed, the basis of a chairman's power within a department was his power within the committee caucus meeting. Failure to carry his pre-meeting meant a loss of face. As one director put it in February 1978: 'We don't want a weak chairman . . . We want someone who's going to carry the pre-meeting with him.'

This gave many chairmen a somewhat competitive attitude towards their colleagues: 'I don't get beat very often' as one subcommittee chairman put it. Certainly, some chairmen seem to have regarded it as their main task to 'get the recommendations through' – to 'marshall the troops', in the words of one councillor – rather than to discuss the reports in an intelligent way. During interview councillors expressed strong feelings about the attitude of some chairmen. For example, a backbench councillor said of one chairman, 'The chairman looks upon himself pretty well as a dictator. It's hopeless because a much wider discussion is required. But everything is cut and dried. The pre-meetings are farcical.' Speaking of a different chairman a main committee vice-chairman said: 'This chairman is a very dominant character . . . He's inclined to force his opinions on people.' This view was supported by a backbench councillor who – speaking of the same chairman – said, 'The chairman doesn't want to hear you. To him he *is* the council.'

The second problem was the duration of pre-meetings. One committee formerly held pre-meetings the night before the main meeting so that there was no risk of running out of time. The chairman mentioned in 1974 that 'sometimes the pre-meetings last longer than the actual committee'. However, he also soon abandoned this practice and other committees only adopted it extremely rarely.

The length of pre-meetings varied from half an hour to one and a half hours and during the period of this study I attended few caucus meetings where there was sufficient time to go into the reports adequately. Meetings were frequently characterised by expressions of impatience. Some Labour councillors were concerned with just getting through the business, whatever the outcome. I measured the length of time spent by three main committee pre-meetings on the agenda items which they discussed. The findings are presented in Table 10.3. Whatever else these meetings were they were clearly not thorough examinations of the issues contained in the officers' reports.

The problem of time had been recognised by the chairmen of two of these committees. In November 1977 the pre-meetings for one (Committee Two in the table) were extended in length to one and a half hours. This did not make any significant difference to the amount of time spent on each item but it did at least mean that most items were reached. Failure even to reach several items had been a frequent

Table 10.3 *Length of Time Devoted to Discussion of Agenda Items by Three Main Committee Pre-meetings*

	Committee One*	Committee Two**	Committee Three***	The Three Committees
Number of agenda items	156	187	275	618
Number of agenda items discussed	149	166	262	577
Time spent discussing agenda items	167 mins	397 mins	586 mins	1150 mins
Average time spent discussing each agenda item	1 min 7 secs	2 mins 24 secs	2 mins 14 secs	2 mins
Agenda items not discussed	7 (4%)	21 (11%)	13 (5%)	41 (7%)

* Four meetings timed, between September and December 1977.
** Six meetings timed, between September 1977 and February 1978.
***Eight meetings timed, between September 1977 and May 1978.

problem when the pre-meetings were only one hour in length. For example, the September and October meetings were one hour in length and had fifty-three items on their agendas. Of these, sixteen (30 per cent) were not reached. The only other pre-meeting for which I kept records was that in April 1977. The meeting was due to be held for one hour. There were thirty-six items on the agenda and the pre-meeting had discussed nineteen of them. That is, seventeen items (47 per cent) were not discussed. This meeting was representative of those held during that slightly earlier period.

It is true that the shortage of time need not have been an insurmountable problem. Many local authorities have a tradition of referring a report back when it is unsatisfactory in some way and detailed work on it is required. There was no tradition of this kind in Newcastle. All the pressure was to get the reports through, even if it needed some detailed attention or was unsatisfactory in some way.

It should not be thought that it was just the chairmen who were the problem. Many committee members simply did not take their jobs seriously enough. I turn now to this problem of the calibre of councillors.

(3) *The Calibre of Labour Councillors*

> Institutions are like fortresses. They must be well designed *and* manned.
> (Popper, 1966, Vol. I, p. 126, his italics)

In discussing some committee caucus meetings I have tried to concentrate wherever possible on the structural or institutional explanations of their conduct, but I have also referred to the individual failings of some councillors. Any adequate discussion of human conduct must take account not only of the structure within which individuals work but also of their personal capacities, values and attitudes.

Concern about the quality of councillors is not a new one. It has been the subject of discussion since the inception of local government in its modern form (Ostrogorski, 1902, Vol. I, p. 490). It was an important issue during the 1960s (Sharpe, L.J., 1962, pp. 201–2), and such criticism is still quite common. For example, this comment was made in the journal *Local Government Review* in 1977:

> One councillor, who retired at the last election with ten years' experience, admitted that he had never spoken in a full council meeting, still did not understand the difference between revenue and capital, and had never read standing orders. There are many like him, and they should not be permitted to serve.
> (Stern, 1977, p. 554)

Many officials have a poor opinion of councillors. For example, in carrying out their study for the Maud Committee, Harrison and Norton found that 'most officers seemed of the opinion that only a minority of members on their committees made any real contribution'. They go on to record some of the descriptions which officers applied to councillors. One said he found it difficult to recall any valuable contributions made by members on committees and subcommittees. Another said that 30 to 40 per cent of his committee were 'useless' (1967, Vol. 5, p. 42). Some officials in Newcastle had a similarly poor opinion of councillors. This is what one senior officer said:

> I am not one who thinks that councillors should be able to go anywhere and to any meeting. For example, there was a meeting this morning on the social services capital programme . . . For this kind of meeting I would not be in favour of councillors just coming in. They would go clogging in, cocking up the process with their size ten boots on occasion. Anyway, they would be completely bored. It is a technical matter for officers – for professionals. It would be an absolute waste of time for officers' meetings to be attended by councillors. If

they weren't getting in the way they probably couldn't contribute and – more likely – would be bored.

Several senior Labour members acknowledged that many back-benchers had little influence but felt that backbenchers only had themselves to blame. As one leading councillor put it, a backbencher's influence 'depends on the contribution of the backbencher. If he just sits quiet, what use is he?' A backbench councillor said, of members of one committee:

Few of the committee members have any interest at all . . . It's easy for the chairman to get through . . . He's got some stooges who will do whatever he asks. Even put their finger in the fire if he asked them to. One backbencher always acts like a bulldozer. She just wants to get through the business.

A subcommittee chairman, speaking of the same committee, said, 'It's very difficult to get a pre-meeting to take a decision against the chairman'. He continued, 'There are too many old women on the committee . . . One of them speaks and then goes to sleep. Another doesn't hear everything that's said . . . A lot of the committee follow the chairman willy-nilly.'

The Maud Committee discusses this question somewhat inconclusively (1967, Vol. 1, pp. 142–4). Other commentators have been less reticent. Jones, in his account of Wolverhampton, cites examples of the criticism of the quality of councillors (1969, pp. 149–62) and there have been many others. Many of these criticisms, however, have been marred by a tendency to associate desirable skills with high economic and social status (for example, Sharp, E., 1962, p. 383; Brennan *et al.*, 1954, p. 82). But as Jones (1969, p. 158), L. J. Sharpe (1962) and Harrison and Norton (Maud Committee, 1967, Vol. 5, p. 41) suggest, this is not necessarily so. The qualities admired by Wiseman (1967, p. 52) – an interest in detail, a capacity for taking pains and common sense – are not confined to any particular social class. This is admirably demonstrated by the career of many of the greatest leaders of the Labour movement.

It is interesting to compare the attitudes of Newcastle Labour councillors with Peter Lee, one of the early Labour councillors. Jack Lawson has described his life as a Labour councillor and miners' leader in County Durham (Lawson, 1936). At first he was a parish councillor and then in 1907 he became a rural district councillor and later a county councillor. Lawson describes Lee's determination to master every detail of his work as completely as he was humanly able.

If new sewers were to be laid down he made a study of the matter. He knew more about the mathematical laws of drainage than most experts in that line . . . If it were a road, the material and all other things received his close attention. Was it water? He calculated the total volume annually as well as the average for each house. And he kept little pocket books full of facts and figures to make sure there was no mistake. They were all reliable, for no book, article or expert gave him a fact or hint which was not recorded. He had all available information ready when he went back to any job he had in hand. He saw it finished in his mind's eye before it started. He knew what he wanted and got it.

This practical, industrious man, driven by a flaming social passion, gripped the people of his district.
(1936, p. 109)

Peter Lee dedicated himself to serving the people of County Durham. To fulfil his duty he felt that he must master every detail of every policy he pursued. He could not know enough. If a problem was complicated he wanted to understand its complexity. If it took time then he gave time. If it meant laborious effort then the effort was made.

The gulf separating such a man from many of the Newcastle councillors is beyond measurement. When faced with complexity or a task requiring either time or effort in one of the caucus meetings, so many Labour councillors contributed nothing. Indeed, their contribution was often negative. If any agenda item led to prolonged discussion it was not uncommon for someone to shout: 'Move next business!' If an issue was complicated they felt uncomfortable. If it required effort they preferred to move on to the next item. If it required time they could not be bothered.

Dennis found a similar phenomenon in his studies of Sunderland. He describes what may happen to a dissident councillor who breaks 'the fundamental rule that what has been accepted as facts in the past and as such been the basis of decisions, are and must remain the facts, and that no amount of scientific evidence can ever disturb that administrative axiom'. He continues:

In these circumstances the uninfluential councillor is an indispensable resource, with his boredom and restlessness backed up as a last resort by the notion that the dissident councillor . . . 'be no further heard' . . . The principle 'the worse, the better' applies, i.e. the more the dissident councillor imagines he has to complain about, the more likely and the more quickly the councillors of the rank-and-file will tire of him and control him.
(1974, p. 168)

A subcommittee vice-chairman in Newcastle described the frustration he felt about the attitude of some of his Labour colleagues:

You've got the opportunity to have plenty of say but whether it has any influence is another matter. It depends on the political awareness of councillors. If they're biased against you it doesn't matter what you say . . . For example, you've got an item up for consideration. It could be bettered. But it's not changed because of the level of the councillors' contributions. They'll not *let you* improve it. So it just goes through . . . It's because of the calibre of the people involved.

I do not wish to give the impression that all the pre-meetings were like this or that all Labour councillors had this sort of attitude, but many meetings were characterised by this sort of behaviour and many councillors' attitudes were of this nature. The principal result was that it was very easy for the chairmen to dominate them:

> But what more oft, in nations grown corrupt,
> And by their vices brought to servitude,
> Than to love bondage more than liberty –
> Bondage with ease than strenuous liberty.
> (Milton, *Samson Agonistes*)

(4) *Party Discipline*

The fourth factor which explained the weakness of backbenchers in the pre-meetings was party discipline. (I have discussed attitudes towards party discipline in more detail in an earlier chapter and here I am only concerned with the implications of party discipline for the pre-meetings.)

Once a decision had been taken by a pre-meeting – however adequate the discussion – all Labour members, whether or not they were present at the pre-meeting, were bound to support it in the main committee meeting. The following statement – made by a Labour councillor – represented the prevailing attitude of Labour councillors to the committee meetings: 'All the arguments should be had in the pre-meetings. The main meeting should be a formality, where the case is publicly put for the press.' Failure to support the pre-meeting decisions in the main committee meetings could lead to disciplinary action being taken. Because the pre-meetings invariably agreed to the reports their principal consequence was, in practice, to increase the power of the chairman by legitimising their use of the party's disciplinary machinery against recalcitrant individuals.

The extent to which strong party discipline weakened the position of Labour backbenchers is demonstrated by the answers some Labour councillors gave to questions 38, 40 and 42 about the power of the opposition backbencher. A number of Labour councillors said that in some ways the opposition members were in a better position than they were because at least they could speak out more freely. For example,

In many respects they are in a better position. They are able to criticise and we are in a jolly difficult position because it's our own party. I can't sit and criticise the chairman's remarks at the main meetings.
(a main committee vice-chairman)

The opposition may stand a better chance than the majority party because they're not governed by the conventions of the party and the pre-meeting. They're more free to use the press and generally to kick up a stink...Party discipline has the effect of committing the majority party backbenchers to the policies of the chairman and the chief officer.
(a main committee vice-chairman)

As the comments made in Chapter 7 show, the view that a councillor must support the majority decision of his colleagues was very strongly held by Labour councillors. This was partly because many councillors identified democracy (which they regarded as an unquestionably good thing) with majority rule. It was also because of the idea of the party as a fighting unit; which meant it was necessary to preserve a united front in the face of the enemy. This led to the strong tradition of following leadership found in some left-wing parties. Michels referred to this: 'The modern party is a fighting organisation in the political sense of the term, and must as such conform to the laws of tactics.' One of these laws was – in the view of Ferdinand Lassalle, the founder of a revolutionary Labour party – that 'the rank and file...must follow their chief blindly, and the whole organisation must be like a hammer in the hands of its president' (Michels, 1968, p. 78). Some Labour councillors in Newcastle were willing to follow Lassalle's advice. As one subcommittee chairman put it, 'There are some committees, if the chairman said "stand on your head" then some of them would'.

Although many Labour councillors acted in accordance with Lassalle's ethic, the prevailing orthodoxy amongst them was strongly democratic in tone. Councillors often stressed that the pre-meetings should take the decisions and that everyone, including chairmen, should be bound by these decisions. Only one Labour councillor – a prominent executive committee member – said during interview that he was in favour of strong leadership. He was a strong believer in the propriety of having very powerful chairmen:

The very fact he's chairman means he's been vested with more authority than anyone else. He's doing most of the work, unlike just the ordinary backbencher. You can't have a lot of little chairmen ... Of course the chairmen have more power. That's why they are chairmen. You choose the chairmen and you give them more power.

Within the Newcastle Labour group most councillors accepted the view that they should abide by the majority decision in pre-meetings on the understanding that there would be fair and rational discussion beforehand. But, in practice, as I have shown above, pre-meetings were merely ritual genuflexions to the principle that full discussion should precede a decision. Few participants took them seriously. Indeed many councillors did not understand the issues well enough to take a serious position on them. Normally, once a chief officer had persuaded his chairman to support a policy that was the end of it. There was little discussion of alternatives. The majority decision meant, in practice, the uncritical – often unthinking – endorsement of over half the Labour members present at any one committee pre-meeting.

Committee chairmen themselves always asserted that they were bound by the decisions of the pre-meetings in the same way as backbenchers. In practice, their relationship with committee Labour groups was rarely of this character. In most cases it was more like the relationship between Roman emperors and the Roman Senate as Gibbon saw it: the emperors 'humbly professed themselves the accountable ministers of the senate, whose supreme decrees they dictated and obeyed' (Gibbon, 1978, p. 38).

Chairmen had very great power. However, just as the group leader was unable to ignore completely some currents of opinion within the party, nor could chairmen totally disregard the views of Labour members of their committee. They had to be able to carry their pre-meetings with them. But, as I have attempted to demonstrate above, the constraints under which the pre-meetings operated meant that, in practice, chairmen enjoyed very wide discretion.

THE INDIVIDUAL LABOUR COUNCILLOR

Chapter 11

THE POWER OF THE INDIVIDUAL COUNCILLOR

The fact that he's a backbench member means that his own colleagues will take very little notice of him. His powers of influence are very small.
(a Newcastle main committee chairman)

How did backbench Labour councillors perceive their own power? If the earlier accounts of the committee process and particularly of the power of the backbencher are accurate, then it is to be expected that a substantial number of councillors would express clear dissatisfaction with the amount of power they had as backbenchers if they were explicitly asked about it. This was done in two ways. Ten Labour councillors were asked questions 56 and 57, and twenty-nine Labour councillors were asked questions 54 and 55.[1] The ten were backbenchers holding no positions or who held only very minor positions.[2] The twenty-nine held positions of at least some consequence.

Of the ten, none thought they had a lot of say, four thought they had some say and six that they had a little say. When they answered question 57, none said that they were very satisfied with the amount of power they had, and three said they were satisfied but not very. One Labour councillor said he had no strong feelings either way and six said they were dissatisfied. Of these six, four said they were very dissatisfied. See Table 11.1.

Table 11.1 *Backbench Councillors: Influence and Satisfaction**

| | | Influence | |
		Some	Little
Satisfaction	Satisfied	1	2
	Dissatisfied	2	4

*Pro-con response excluded.

The two councillors who said they had little power but were satisfied felt this was reasonable because they were inexperienced. For example, one said:

> I sometimes feel I'm banging my head against a brick wall. A lot of officers regard councillors as a pain in the side . . . With being a new councillor I have to accept it. With more experience I would not be satisfied.
> (a backbench councillor)

This is what councillors who felt they had little power and who were dissatisfied said:

> When I first started here I had nowt, but it got better after a time. When they get to know you it gets better . . . Officers can get than damn powerful, they won't talk to you.
> (a backbench councillor)

A second councillor gave an example of the disrespect which characterised the attitude of some officers:

> Go to any office and say 'I'm a councillor.' You might as well say you're King George. It doesn't mean a thing to them. I went to social services, to the inquiry counter, and this officer said, 'OK, just get in the queue behind that woman'. He knew I was a councillor because I'd had a battle with him before. I should have got on the phone to him from the members' services and said: 'Get down here quick'.

This is how one chief officer justified the low esteem in which many officials held councillors: 'A lot of you bring it on yourselves', he said. 'I get, say, three phone calls a day from [a named councillor], one asking for someone to call on Mrs Hunter of Davis Street and it turns out to be Mrs Davis of Hunter Street; that sort of thing.'

The tendency to accord backbench councillors low status, or worse, was not confined to Newcastle. Roland Freeman, for example, reports that at 'London's County Hall . . . the ordinary member is generally regarded as a rather low form of life' (Freeman, 1977b, p. 441). In Sunderland, Dennis found that 'In private the officials . . . expressed only contempt for councillors. Not only did the councillors know nothing . . . they were not entitled to know anything' (1972, p. 238).

Councillors who were chairmen or vice-chairmen of either main committees or important subcommittees were asked to distinguish between their roles as chairmen and vice-chairmen and their role as a backbencher. They were asked questions 54 and 55. Twenty-eight councillors answered these questions. Three (11 per cent) thought they had a lot of say as backbenchers, fifteen (54 per cent) thought they had some say, eight (29 per cent) thought they had little say and two (7 per cent) thought they had none.

None were very satisfied with their power as backbenchers though twelve (43 per cent) were satisfied. Four (14 per cent) had no strong feelings either way and twelve (43 per cent) were dissatisfied. Of these twelve, five councillors (18 per cent of the total) were very dissatisfied. See table 11.2.

Table 11.2   *Chairmen and Vice-chairmen, in their Capacities as Backbenchers: Influence and Satisfaction**

		Influence					
		Lot	Some	Sub Total	Little	None	Sub Total
Satisfaction	Satisfied	2	9	11	1	0	1
	Dissatisfied	1	3	4	6	2	8

Not ascertained, 1.
*Pro-con responses excluded.

The following are some of the statements made by Labour councillors in answering these questions:

It depends on what I say. If it warrants it, a lot [of influence]. I've often got things accepted or delayed. Seniority helps. Experience. (a subcommittee chairman who said she had a lot of influence as a backbencher)

I've never felt my time was completely wasted. If you make constructive remarks it seems to be listened to. (a main committee vice-chairman who said he had some influence as a backbencher)

If you're chairman of one patch you tend to keep out of the others. If I was purely a backbench member then I could understand why people come and go. As a backbencher you are pure lobby fodder . . . You get the reports two or three days before the committee. You can't be expected to put forward serious alternatives. You're on a hiding to nothing. (a main committee chairman who said he had little influence as a backbencher)

The whole process is a nonsense. Information is withheld from me. When you get it, it's given in such a way that you find it difficult to use . . . I would apply this to all committees. Everything is sorted out before it's started so there is only one line to be taken . . . I've always felt powerless. If I didn't go to any meetings it wouldn't worry me. Democracy is a myth. It doesn't exist. There are thousands of examples. (a main committee chairman who said he had no influence as a backbencher)

The findings from questions 54, 55, 56 and 57 are summarised in Table 11.3.

Table 11.3 *Councillors as Backbenchers: Influence and Satisfaction*

	Chairmen as backbenchers Column (a)	Backbenchers Column (b)	Column (a) plus Column (b)	
Power:				
Lot	3	0	3 (8%)	
Some	15	4	19 (50%)	
Little	8	6	14 (37%))	16(42%)
None	2	0	2 (5%))	
Total	28	10	38 (100%)	
Satisfaction:				
Very satisfied	0	0	0	
Satisfied	12	3	15 (39%)	
Pro-con	4	1	5 (13%)	
Dissatisfied	7	2	9 (24%))	18(47%)
Very dissatisfied	5	4	9 (24%))	
Total	28	10	38 (100%)	

Not ascertained, 2.

These findings are further borne out by the answers to questions 58, 59 and 60 about the councillor's power as a non-member of a committee.

Thirty-eight councillors answered 'Yes' to question 58 and two answered 'No'. Of the thirty-eight, thirty-one answered question 60. No one thought that success was very likely. Only six councillors (19 per cent) thought it was fairly likely and the vast majority, twenty-five councillors (81 per cent), thought it was not likely.

Thus a total of twenty-seven councillors thought that it was either 'not likely' that they would be successful or that there was nothing they could do. The following are examples of comments made whilst answering this question:

It depends. Some chairmen think the officers can't do wrong. I've seen chairmen speaking with the officer's voice and not that of the party. (a main committee chairman)

It's very unlikely. You would be treated as an outsider. (a backbench councillor)

Some chairmen make their minds up first and if anyone disagrees they take it as a personal matter. (a main committee chairman)

One problem for councillors was that it was in practice unlikely that they would hear about a recommendation until after the meeting anyway. (Although about a year after the interviews this was changed.) Some councillors made this point during interview. For example: 'The problem is getting to know about it in the first place. You would only find out if you were told about it by some one who was very aggrieved' (a main committee chairman).

These were some of the frustrations felt by backbench Labour councillors. Many were dissatisfied with the amount of influence which they were able to exert. A major frustration was with the low esteem in which they were held by some officers and some leading councillors. This was important, not because of the personal slight involved, but because councillors simultaneously faced an electorate which expected a great deal of them. Individual citizens contacted them by telephone or letter or by calling at surgeries and expected them to act effectively on their behalf: to get a council house repair done, or to secure a transfer from one council house to another or to put right a grievance. Tenants' or residents' associations looked to councillors to act on their behalf, demanding perhaps a new community centre or a bigger grant or an improved playscheme. A councillor might find himself being held responsible in an angry public meeting for a policy on which he had had scant influence. For councillors with a keen sense of their duty to the electorate the sense of frustration produced by these conflicting pressures was particularly acute.

Having described the feelings of Labour councillors I turn now to a discussion of two specific problems which they faced: the limited time and the limited information at their disposal.

NOTES: CHAPTER 11

1 The lord mayor was excluded.
2 Five held no official positions and five were vice-chairmen of subcommittees with only very limited powers. (Not ascertained, one.)

Chapter 12

THE COUNCILLOR'S RESOURCES

> This superiority of the professional insider every bureaucracy seeks
> further to increase through the means of *keeping secret* its know-
> ledge and intentions. Bureaucratic administration always tends to
> exclude the public, to hide its knowledge and action from criticism
> as well as it can . . . In facing a parliament, the bureaucracy fights,
> out of a sure power instinct, every one of that institution's attempts
> to gain through its own means . . . expert knowledge from the inter-
> ested parties. Bureaucracy naturally prefers a poorly informed,
> and hence powerless, parliament – at least insofar as this ignorance
> is compatible with the bureaucracy's own interests.
> (Max Weber, 1968, Vol. 3, pp. 992–3, his italics)

In this chapter I discuss two resources crucial to the councillor, those
of time and information. At the end of the chapter I consider some
recent improvements in the opportunities available to Newcastle's
councillors.

THE PROBLEM OF TIME

How much time did Labour councillors spend on their council duties
and which activities took up most of it? The Robinson Committee
found that the total amount of time spent on council work had
increased by about half from fifty-two hours per month in 1964 to
seventy-nine hours per month in 1976. The figure was higher for urban
authorities. The average time per month spent by councillors on council
duties, in the English metropolitan districts in 1976, was 109 hours.
How do these figures compare with those for the Newcastle Labour
group?

Labour councillors were asked to say how many hours they spent on
council work during the previous week. When they felt the previous
week had been strongly unrepresentative they gave an estimate for the
nearest more representative week. Thirty-nine Labour councillors
answered this question. Six said they spent over forty hours a week,
eleven said they spent thirty-one to forty hours, thirteen said they spent

twenty-one to thirty hours, eight said they spent eleven to twenty hours and one councillor said he spent less than ten hours. Thus, thirty councillors (77 per cent) claimed to spend twenty-one hours a week or more on council work.

Labour councillors were asked to say how much time they spent on particular activities – attending meetings, reading reports, and so on – but their estimates were not very reliable. I therefore recorded the time which I spent myself on activities associated with council work over a seven week period. The results are set out in Table 12.1.

Table 12.1 *Time Spent on Council Activities by One Councillor*

		Seven-week total	Weekly average
1	Reading committee reports	26 hrs 27 mins	3 hrs 46 mins
2	Correspondence, telephone calls, etc.	32 hrs 40 mins	4 hrs 40 mins
3	Attending meetings as a backbencher	48 hrs 9 mins	6 hrs 52 mins
4	Attending meetings in capacity as chairman	16 hrs 43 mins	2 hrs 23 mins
5	Labour party meetings	12 hrs 21 mins	1 hr 45 mins
6	Committee pre-meetings and group meetings	20 hrs 29 mins	2 hrs 55 mins
7	Surgeries	9 hrs 15 mins	1 hr 19 mins

Items 3 – 7 include travelling time.

Over the seven-week period I spent a total of 153 hours forty-three minutes on council duties, at a weekly average of 21 hours fifty-eight minutes. For the purpose of comparison with the findings of Maud and Robinson I have calculated a figure for one month; it is based on the daily average achieved over the seven-week period multiplied by thirty. The figure for one month was 94 hours seven minutes. This was above the average for Great Britain but below that for metropolitan districts in England.

The problem of simply keeping up with the reading was a serious one because of the large volume of paper sent to councillors. I counted how much I was sent over a six-month period from September 1977 to February 1978. During this period I received, mostly unsolicited, 4,431 printed sides of A4 paper in the form of reports or other documents. In addition, I received 382 letters; most were from council departments, but forty-five letters were from constituents. Thus, I received in a week an average of 170 sides of A4 paper, thirteen official letters, and just under two letters from constituents. During the seven-week period reading committee reports took an average of three hours forty-six minutes a week. This may seem rather low, given the volume of paper received. This was because I normally scanned papers in the first instance and only read carefully those that had important implications. It would have been useful to read them all thoroughly but the shortage of time made this impossible.

In addition to reading the reports, dealing with correspondence also consumed time. I counted the number of letters I wrote during my first full year on the council (from 1 January 1976 to 31 December 1976): 1,197 letters were written at a weekly average of twenty-three. During the seven-week period an average of four hours forty minutes per week was spent dealing with correspondence and making telephone calls.

In addition to being asked how many hours they spent on their council activities Labour councillors were also asked to say which of their council activities took up most of their time (question 25). The results can be found in Table 12.2.

Table 12.2 *Activity Which Takes Up Most of the Labour Councillor's Time*

Activity	Number of respondents	%
Work in the ward	9	24
Job as:		
leader/lord mayor/ chairman/vice-chairman	13	35
Meetings/committee work	10	27
50/50 ward work and committee work	5	14
Total	37	100

Not ascertained, 3.

Labour councillors were also asked to say whether or not they were able to give all aspects of their work on the council enough attention because of the problem of time (question 26). Thirty-eight Labour councillors answered this question. Twenty-seven (71 per cent) said they were unable to give some aspects of their work enough attention and eleven councillors (29 per cent) said they were able to give all aspects of their work enough attention.

Councillors who found time a problem were then asked to say what aspect of their work suffered most (question 27). Twenty-seven councillors answered the question. Eleven (41 per cent) said their ward work suffered. Six (22 per cent) said that they had difficulty reading all the reports they were sent and the others made a number of different points.

For example, one councillor felt that his job as a vice-chairman of a main committee suffered. 'I don't really know some parts of the town well enough . . . You need full-time councillors.'

One main committee chairman felt that both his job as chairman of a main committee and his ward work suffered: 'Constituency work suffers very heavily. As chairman, policy research work [suffers]. Getting to . . . see the system in operation. I tend to move in ever decreasing circles within the office. The problem is getting everything in.'

THE PROBLEM OF INFORMATION

Daniel Bell, in *The Coming of Post-Industrial Society*, draws attention to the growing importance of knowledge and technology in modern societies: 'the two major axes of stratification in Western society are', he says, 'property and knowledge' (1973, p. 43). Local government in this country is characterised by growing complexity, to the extent that it is increasingly difficult for councillors to keep themselves well informed. The reorganisation of local government did not improve matters. Soon after local government reorganisation, J. D. Stewart warned about the dangers of increased centralisation which the reorganisation had encouraged. 'Councillors outside the central group' should, he argued, 'receive support through the development of new forms of councillors' information services.' He warned that unless this and other steps were taken, 'the conflict between a centralisation of political power in the local authority and a desire by more councillors for active involvement . . . [might] endanger some of the newly created organisational structures' (Stewart, 1974b, p. 32; see also 1974a, pp. 118–21).

In the questionnaire councillors were asked two questions about their access to official information:

Has anyone ever put difficulties in the way of your getting access to information which you needed to carry out your duties as a councillor? (question 61)

Councillors who answered positively were then asked:

Has this happened *often* or *on some occasions but not often* or has it happened *only rarely?* (question 62)

Thirty-seven councillors answered question 61. Of these, sixteen (43 per cent) replied 'Yes', and twenty-one (57 per cent) replied 'No'. Of the sixteen who replied positively, four (25 per cent) said it happened often, eight (50 per cent) said it happened sometimes and four councillors (25 per cent) said it happened only rarely.

Councillors who said it happened often had this to say:

It happens *very* often.
(a main committee chairman)

It's only partly deliberate. Often, just as a matter of routine, they don't believe in giving information out. They don't believe in the propriety of giving councillors information which they need.
(a main committee vice-chairman)

Councillors who said it happened sometimes said this:

[A named department's] big get-out is 'client confidentiality'. It's hard to battle against. [A named area office] refused me a report on an item. I explained I was chairman, very kindly, and it made no difference . . . Eventually I got it through the director.
(a main committee chairman)

I still have difficulties as chairman. They don't tell you things which they fear you may not like . . . They do it by giving you something other than what you asked for, or by being too slow.
(a main committee chairman)

Some councillors mentioned that the problem was much worse in opposition: 'If you're in opposition some officers are most autocratic', as one advisory committee chairman put it.

Of the councillors who answered negatively, one subcommittee vice-chairman said it might be because of the nature of his requests for information: 'It might be because of my age. I can approach them so it's impossible for them to refuse. Maybe I have only made moderate requests for information. Maybe the information I wanted hasn't been all that important.' A backbench councillor said he didn't ask for enough information to be able to answer the question: 'I'm not sufficiently active in asking for information. When I was younger, yes. I was more active, but not so now. So I could not give you a reasonable answer.'

This councillor draws attention to an important point. How valid is his suggestion – that a councillor's answer to this question (question 61) will depend upon the extent and quality of his activities as a councillor? The above backbench councillor's view is partly supported by the answers which Labour councillors gave to the question asking them how much time they spent on council work.

Councillors who said that they had been denied information claimed to spend more time on council work than those who did not. Nine out of the fifteen councillors (60 per cent) who answered 'Yes' to question 61 said they spent thirty-one hours or more a week on council work. Only

33 per cent of those who answered 'No' made a similar claim. Fourteen out of the fifteen councillors who answered 'Yes' (93 per cent) claimed to spend twenty-one hours a week or more on council work while only 62 per cent of those who answered 'No' did so. Of the eleven councillors who said they were refused information often or sometimes, seven (64 per cent) claimed to spend thirty-one hours or more a week on council work.

COUNCILLORS' ATTITUDES TOWARDS ACCESS TO INFORMATION

Labour councillors were also asked to say how extensive they thought a city councillor's access to information should be. They were asked to say whether they agreed or disagreed with the following statements:

All councillors *in the majority party only* should have a right to any information in the possession of the council. (question 80)

All councillors *regardless of party* should have a right to any information in the possession of the council. (question 81)

According to their answers to both of these questions Labour councillors have been classified as follows. First, there were those who felt that majority party councillors only should have a right to information in the possession of the council. Ten councillors (25 per cent) took this view. Secondly, there were those who felt that all councillors should have a right to information. There were nineteen (47 per cent) such councillors. Thirdly, there were those who felt that no councillors should have a right to any information. Eleven (27 per cent) councillors took this view. What follows are some of the statements made by councillors in each of these categories.

Majority party only:

The Tories should not have *any* information. When they are in power it's the other way round. They don't give us information.
(a backbench councillor)

We may not want the Tories to know about something. They may spoil it in the meantime. I wouldn't tell them anything (laughing).
(a subcommittee chairman)

All councillors:

This can be a difficult one, but if it's available for one it should be available for all.
(a main committee chairman)

Two Labour councillors in this category expressed reservations. For example, a main committee vice-chairman said:

> If you're in opposition you're just wasting your time. If it was left to me I wouldn't give the Tories too much. I agree they should get it but if we were in power for ever that would be a different matter.

No councillors:

> You can't run the council if you allow papers out prematurely.
> (a main committee chairman)

One backbench councillor disagreed with both statements on the ground that some councillors did not read the information which they already received:

> There is some information, why should I want to know about it? What the hell would you want it for? Some of the information some councillors get now, they just throw it out. There is too much just left lying about. They just leave their reports behind after the meeting. Half of the buggers never read the reports anyway, therefore I wouldn't think they needed more information. They don't read what they've got now. You shouldn't have every little thing that's going on. What about the cost? It's not necessary. There's plenty of time to see it when it comes to be released anyway.

Thus a majority of Labour councillors had restrictive attitudes towards information. To some extent they were themselves the victims of these attitudes, as the following case studies show.

THE COUNCILLOR'S ACCESS TO INFORMATION: SOME CASE STUDIES

From the above account the way in which the councillor's access to information was problematic begins to emerge. In order to portray more fully the extent and character of the problem, I have set out below two case studies of the degree of access to information enjoyed by council members.[1] Before turning to the case studies it is worth dealing with a common objection to this type of material. It may be argued that the events described below only took place because of the attitude or behaviour of the present author, and that therefore the findings can be discounted as unique or at least exceptional. The main flaw in this assertion is that it grants to the present author an importance in the affairs of Newcastle City Council which he did not have. Did the prominent officers who feature in the case studies behave towards the councillor for Blakelaw ward in a manner which was not only different from their normal behaviour but so exceptional that the case study may be discounted as a valid observation of official behaviour? On the contrary, there is no reason to believe that these powerful officers behaved in a unique or exceptional fashion. Having said that the findings cannot be dismissed is not to assert that they are somehow more than single observations. All that is shown is that in the cases studied a councillor seeking to obtain specified information found difficulties put in his way.

Council House Modernisation
Immediately after being elected, the Councillor began to take up issues which had been raised during the election campaign. One such issue was that of council house modernisation; many electors wanted to know whether or not their homes were going to be modernised. The Councillor set out to discover whether any parts of his ward were in the council's modernisation programme. He eventually discovered that the bulk of it was not. He therefore set out to discover whether or not there were any special problems which would make it difficult to put the particular estates into the programme; for example, were central government guidelines a problem or was there a local shortage of cash? It transpired that the information which the Councillor needed was contained in four documents:

(1)   The modernisation programme.
(2)   The *Manual on Local Authority Housing Subsidies and Accounting*.
(3)   A circular letter from the Department of the Environment on 'Delegated cost limits for subsidy purposes'.
(4)   A letter from the Department of the Environment allocating Newcastle a special sum to be spent on modernisation to 'relieve unemployment in the construction industry'.

The Councillor first approached the relevant area housing manager, who said that the central government did not allow local authorities to spend money at all on the modernisation of postwar council houses. The Councillor decided to check and arranged to see one of the assistant directors of housing.[2] He asked him for the most recent government instructions on the modernisation of postwar dwellings. The Assistant Director of Housing replied that they were contained in two documents, the *Manual on Local Authority Housing Subsidies and Accounting* (document 2, above) and in a recent letter setting out delegated cost limits (document 3, above).

'Do they rule out the possibility of modernising postwar council housing?', the Councillor asked. The Assistant Director of Housing replied that he thought they did, but added that he had not had time to study the documents fully and so he could not be sure; one document was very complex. He then went on to describe their contents somewhat vaguely.

'Yes, well, I obviously need to be absolutely certain. I can't just say that we are not sure. Could I perhaps have copies of those documents to study myself?', asked the Councillor.

The Assistant Director of Housing replied that he had only got one copy of the *Manual* and that he was using it.

'Could I borrow it when you are not using it?', the Councillor asked.

'I'm using it all the time.'

'*All* the time?'

'Yes, *all* the time.'

'OK, well I'll try to get it from someone else, but what about the letter you mentioned, could you not photocopy that for me?'

The Assistant Director of Housing replied, 'Ah, that is addressed to the Chief Executive. I'm afraid I cannot give you a copy of a personal letter.'

'It's not personal in the usual sense, it is addressed to him only in his capacity as head of the council's staff', said the Councillor.

'I can't let you see it.'

Well, why don't you ring up the Chief Executive and ask him whether he minds? I'm sure he won't.'

In spite of having two on his own desk, the Assistant Director of Housing went out of his room to use a telephone. He returned to say that he had been unable to contact the Chief Executive but that instead he had asked the Director of Administration. He had ruled that the letter should not be released. The Councillor asked the Director of Administration why he had taken this decision and in a subsequent letter he gave the following reason:

> In general terms a councillor is entitled to sight of documents only for bona fide purposes connected with his work on the council: each case has to be considered on its merits and I am quite unable to decide whether or not you should have a copy of a particular letter without the opportunity of seeing what it is and asking why it is you want it.

In the same letter he announced that he had agreed to release the document in question because it was available to members of the public anyway. Thus, the Councillor had met with success in the particular case. But, the Director's letter seemed to raise a more important general issue. It held out the prospect of a lengthy wrangle on every future occasion on which the Councillor sought to acquire information. He therefore took up the question with the Chief Executive who announced that he was about to circularise chief officers with his guidance on the provision of information for members. This circular started well, but went on to list exceptions:

> It is, in my view, of the utmost importance that he [a councillor] should be given as much help as possible in order to enable him to carry out [his] various roles as effectively as possible and I look to everyone concerned to assist members in the achievement of this objective.

> There are a limited number of occasions where, in the chief officer's judgement, disclosure of information to an individual member of the council *may be premature or may have potential policy or political consequences*, then the chief officer would wish to refer the matter to the chairman of the appropriate committee.
> (my italics)

Information, the disclosure of which might be deemed 'premature' or which might have 'potential policy or political consequences', could hardly be described as limited in quantity. Furthermore, the only safeguard being proposed was that the chief officer should consult the appropriate committee chairman. Normally, because of the nature of the committee process and the importance and closeness of the chief officer—chairman relationship, it did not prove difficult to enlist his or her support. These guidelines, then, could have been used to justify any amount of secrecy.

Meanwhile the Councillor continued to try to acquire other information. He needed to know what the present modernisation programme was and again approached the Assistant Director of Housing, who replied that he would have to check with the Director of Housing that it was in order to release the document. He agreed to let the Councillor know what the Director of Housing said.

He did not do so and the Councillor rang to remind him. He was told that the Director of Housing had decided that the document was not to be released. Furthermore, the director had asked the Assistant Director of Housing to find out why the Councillor wanted it. The Councillor said he would speak to the Director of Housing himself and, after a couple of days of trying to get him on the telephone, the Councillor asked for a copy of the modernisation programme.

'It has only just been submitted to the Department of the Environment and they have not yet approved it', said the Director of Housing.

'Fair enough, I appreciate that the Department of the Environment have not approved it, but the point is that this council has agreed to submit the document to them and what I want to know is what exactly we, that's this council, have agreed. I realise the Department of the Environment might turn down some of it.'

'Well, I don't know; why do you want it anyway?', said the director.

'To see if my ward is included.'

'Is that the only reason?'

'Yes.'

'Well, if that is the only reason I can understand that.' He agreed at the time to have a copy put in the post. He did not do so. Whether he decided subsequently that there must after all be other reasons is not clear. What is clear is that the Director of Housing thought it was up to him to decide.

At about this time the Councillor heard a rumour that additional funds had been allocated to Newcastle to relieve unemployment in the construction industry. Could this be spent in his ward? He again approached the Assistant Director of Housing who said he could not discuss the matter; the Councillor must see the Director of Housing. The Councillor arranged an appointment for a few days later when the following conversation took place:[3]

Councillor: I understand there is some extra money for modernisation associated with the job creation programme. Can you tell me a bit about ?

Director: Well, there's about half a million pounds. Certain conditions are attached, like, it has to be labour-intensive, approval for schemes has to be sought; it's a bit like Section 105 in that sense, plus one or two other things.

Councillor: So it's fairly complicated then. Was this a circular or a letter or something else?

Director: A letter.

Councillor: Oh, well could I have a copy of it so I can study it more carefully?

Director: Er . . . no.

Councillor: No! Why is that?

Director: Well, I just don't think it's necessary.

Councillor: This seems extraordinary to me. Here we have a member of the council, a corporate body with various powers and which among other things employs you, and you are saying to the member that he can't see a letter which has been addressed to that council.

Director: Well, why do you want it? I mean, what if all councillors asked for this sort of information? You represent Blakelaw ward, if I let you have it, it will apply to everyone else.

Councillor: Of course, but that is not a problem at the moment; there is only me that wants it.

Director: Yes, but why?

Councillor: Because I want to inform myself, so that I can make a contribution to any debate which takes place on modernisation. Also, of course, I want to see if some of it can be spent in my ward.

Director: What debate?

Councillor: The discussion that takes place on this policy, about how and where to spend this money.

Director: But that will take place in the Housing Committee. There will be a report on it . . .

Councillor: Yes, but I am interested in being informed about this and putting forward a point of view at this stage. It's a very important matter and I expect the group will discuss it. If I wait till you prepare a report to the committee, that will be too late – it's cut and dried by then.

The Director of Housing eventually said he would put a copy in the post. He did not do so and the Councillor rang to remind him. The Director then said that, because the chairman had not seen it, he had on reflection decided not to release it.

'Well, send him a copy too', suggested the Councillor.

'It has not been to committee yet, I will have to ask the chairman.'

The Councillor approached the chairman of the Housing Renewal Committee, Councillor Arthur Stabler, and he agreed to speak to the Director of Housing. Two days later the Councillor rang the chairman.

'Have you seen the Director of Housing yet?'

'Yes.'

'What was the outcome?'

'You already know what the document says,' replied the chairman. 'And anyway, if I let you have it, anyone could get it, including the Tories.'

'The Director of Housing tried to explain some parts of it, yes, but I need to study it myself because he did not know the answer to the particular point I was interested in.[4] On the question of the Tories, that is irrelevant, there is nothing controversial about this, we have simply been given more money,' replied the Councillor.

'No, I have discussed this with the leader and he agrees that we should not establish the principle that information should be available to any councillor because that would include the Tories. You are trying to establish a general principle of a councillor's rights to information but that would include the Tories. I remember what the old town clerk used to say when I first came on to the council: "You're a councillor in the council and you're a councillor in committee but outside them you're just plain mister."'

'I'm sure he did say that, but he is wrong. Councillors have a clear right to information which they need to carry out their duties,' replied the Councillor.

'But that would include the Tories. The leader agrees that we are not going to accept the principle,' retorted the chairman.

'One logic of that statement is that you will gladly give me the information as a party comrade, but not as a councillor. Is that so?'

'Ah,' said the chairman, 'so you only want it for yourself. You are always on about open government, when all you really want is information for yourself. You are not bothered about the principle.'

'And you are, I take it', said the Councillor.

'Yes, I've always believed in open government.'

'Let me have the document then.'

'No, the leader has decided you can't have it. That would mean the Tories could see it.'

The Councillor then went to see the leader. He agreed to let the Councillor have the letter as a personal favour; the Director of Housing would be told to release it. However, in fact the letter was never sent. Reminders to the Director of Housing and the leader's secretary produced no response. Eventually the Councillor gave up. Of the four documents which the Councillor had sought, three were eventually obtained (documents 1, 2 and 3). The fourth was not.

What legal rights does the councillor have to information? The basic principle was succinctly stated in *R. v. Barnes Borough Council ex parte Conlan* [1938].[5] It was held that a councillor had a right to inspect all documents in the possession of the council 'so far as his access to the documents is reasonably necessary to enable the councillor properly to perform his duties as a member of the council'. The judge said:

The common law right of a councillor to inspect documents in the possession of the council arises from his common law duty to keep himself informed of all matters necessary to enable him properly to discharge his duty as a councillor.

This right is however, qualified. The principal qualifications are set out in the following two cases. In 1907 it was held that 'a member of a municipal corporation has no right to a roving commission to examine the books or documents of the corporation merely because he is a member of such corporation' (*R. v. Southwold Corporation ex parte Wrightson* [1907].[6] Secondly, the courts have refused to compel disclosure if they felt that a councillor was not motivated solely by his public position. For example, in the case of *R. v. Hampstead Borough Council ex parte Woodward* [1917][7] it was found that a councillor wished to assist a third party in litigation with the council and for this reason the court refused to compel release of the documents.

In addition to the common-law right to information, as qualified, the councillor also has some statutory rights of access to documents, most of which are held in common with all local government electors or persons interested. These are set out below.

Section 159 of the Local Government Act 1972 gives a right to any person interested, at each annual audit of a local authority's accounts, to inspect and make copies of the accounts to be audited and all books, deeds, contracts, bills, vouchers and receipts relating to the accounts. In addition, any local government elector can question the auditor about the accounts. A further power is available in Section 228 of the same Act which entitles a local government elector to inspect and make copies of council minutes, any order for the payment of money made by the council and the abstract of accounts of the council, or of any officer or any auditor's report on the accounts. That section also gives an exclusive right to a member of a council to inspect the actual accounts of the council or the accounts of any proper officer of the council.

I turn now to a discussion of the Director of Administration's letter (of December 1975) quoted above (p. 150). He put forward a particular interpretation of the councillor's legal rights. He appeared to reserve the right to see any document which a councillor requested and to withhold it if he was not satisfied with the reason(s) for the councillor's wish to see it.

The Councillor tried to clarify this point later, in 1976 and in early 1977. The following correspondence took place. The Councillor asked:

> Do you, as a general rule, assume that members want information in order to carry out their duty or do you assume, as a general rule, that they have some other motive(s) and that, therefore, you will deny them access to documents? My own view is that the normal presumption should be that documents are to be made available to members unless the chief officer concerned can show that a councillor wants the information for some reason other than his wish to carry out his duty. I have a faint suspicion that this is not the general rule upon which you and other officers act and would be grateful if you could set out just what present practice is.

The reply confirmed the Councillor's fears. The burden of proof was on a councillor to justify disclosure and not on an official to justify withholding information:

> As I see it the common law rule entitles a council member to information and documents where he can demonstrate that he has an interest in receiving the information and that his motive for asking for the information is consistent with the interests of the council as a whole.
>
> I would therefore regard it as the duty of the councillor to make known the reasons for his requesting information to the officer in question and if there is any doubt in the officer's mind he may need to question the member further. If he remains under any doubt about whether he is under some duty to disclose the information then in the

last resort the request would have to be put before the appropriate council committee.

The Education Department

In March 1979 the Councillor set out to discover some basic information about the city's schools to enable him to reach a point of view about the success of the city's educational policies. He wrote to the Director of Education asking him for the following information:

(1)   Comprehensive school examination results over a four-year period.
(2)   The percentage of each comprehensive school's intake which required remedial treatment.
(3)   The percentage fall-out rate of sixth formers in each school.

The Director replied as follows:

Thank you for your letter . . .
(1)   *Examination results of comprehensive schools*:
       These are available for each school to the members of the Governing Body of that school, but we have never provided them for every school even to members of the Education Committee since comparisons are misleading.
              This has been a long-standing practice over more than ten years. You will find figures for the authority as a whole compared with the national figures in 'The Quality of the Education Service'.[8]
(2)   *The remedial service*:
       Staff have working papers giving this information but it is for use with schools and teachers only.
(3)   *The fall out of sixth form numbers*:
       We have working papers on this but the matter has not yet been pursued even by the Education Structure Group.
              I regret, therefore, that I am not in a position to let you have the information you ask for. I am letting the Chairman of the Education Committee have a copy of this reply.
                            Yours sincerely,
                            (signed)
                     Director of Education

The committee chairman supported the decision of the director. She initially gave two reasons for doing so. First, she refused to release the information to the Councillor because if she did so then it would also have to be given to opposition councillors who might abuse it. Secondly, she argued that the information had not been released throughout the previous ten years and that therefore it should still not be released. The Councillor raised the issue with the group leader and as a result a special meeting of group officers was arranged to discuss the issue. It was held on 9 April 1979. At this meeting the chairman of the Education Committee and other supporters of her view expanded on their reasoning.

The meeting began with demands that the Councillor should say why he wanted the information. The chairman and supporters then advanced the following arguments. They chose to argue as if the only information being sought was the examination results.

(1) They argued that any comparisons between schools would be misleading.
(2) They argued that if you compare the examination results of schools you will encourage people to believe that they are the only things that matter.
(3) They said that the information would be used by opponents of comprehensive education, that it would be putting tools in the hands of an unscrupulous press.
(4) They argued that a league table (which had not been requested) would be damaging to some schools.

Nevertheless, at this meeting of group officers a compromise was reached. The councillor who was making the request was to be provided with the information in confidence. Secondly, ways in which examination results could be released would be explored, but that in the meantime they would not be released. Quite coincidentally, a resolution along these lines was agreed by the Labour group as an amendment to an opposition motion submitted to the council meeting in June 1979. It read as follows:

This council directs the Education Committee through its officers to seek ways in which more information about schools can, in the light of various factors which are relevant, be made available to all councillors without creating invidious comparisons or league tables.

Whatever view is taken about the release of examination results this should not be allowed to confuse the basic issue. Certainly, as the chairman was anxious to point out, there is more to education than examination results alone. This was not the issue. Examination results are one important measure of a school's performance. No more was being asserted by the advocates of their release.

The issue was this. The authority spent many millions on education. It consumed over half its revenue budget. The secrecy over the examination results, the fall-out rate and the usage of the remedial service meant that councillors were unable to judge how effectively it was being spent. The council had a policy of discriminating in favour of schools in deprived areas. Was this policy working? Councillors were unable to answer such questions. The Education Department were able to deny councillors the means by which they could rationally assess the success of the department.

The net result of the secrecy in Newcastle City Council was, as one senior officer put it, 'There must be only seven or eight people in the whole council who know what's going on. They just close ranks when you ask for information.' Some readers may not find the difficulties experienced in the case studies all that disturbing. They may feel that, after all, many of the documents in question were eventually obtained, even if after considerable delay.

However, such an attitude could only be attractive to someone who was unaware of the wider circumstances in which such struggles took place. The sort of difficulty described above could have a crippling impact upon the councillor's effectiveness, because the councillor has only limited resources of time and energy. Most councillors had full-time jobs. Their routine council commitments were demanding and councillors who took their role seriously found great difficulty in fulfilling these ordinary obligations. If the councillor was compelled to spend a lot of time on any particular activity (for example trying to obtain access to known documents as in the above cases) the time available for other duties was correspondingly reduced. A main committee vice-chairman drew attention to this problem. He mentioned the last occasion on which he had been refused information: 'I was told lies and could not get further information.' But, he continued, 'I had to drop it in the end because I had no time to pursue it. Other things were being affected.' For councillors, time was an important political resource. If their time was wasted their effectiveness could be reduced.

There is another sense in which time could be an important factor. Councillors normally needed particular information by a specific date (a committee meeting, say). This meant that the timing of the release of information was also important. (I have discussed this issue in more detail, above.) The nearer to the meeting that information was released the weaker the position of the councillor; and better still, from the point of view of power-holders, if information could be withheld until after the time of the effective decision then the councillor could be completely emasculated.

For these reasons information was a key weapon in the daily struggle that took place for influence within the local authority. A councillor wishing to test the truth of a statement or the validity of a chain of reasoning needed information. Secrecy enabled officials to protect their statements from factual or logical refutation. Max Weber was particularly critical of the tendency for officials to be secretive in order to avoid scrutiny of their activities. He argues that apart from being rooted in the administrative division of labour the power of all bureaucrats rests upon knowledge of two kinds: 'First, technical know-how in the widest sense of the word, acquired through specialised training.' This expertise alone, he says, does not explain the power of the bureaucrat:

> In addition, the bureaucrat has *official information* which is only available through administrative channels and which provides him with the facts on which he can base his actions [his italics]. Only he who can get access to these facts *independently of the officials' good-will* can effectively supervise the administration [my italics].
> (Weber, 1968, Vol. 3, pp. 1417–8)

IMPROVEMENTS

During the main committee process there was no setting in which the ward councillor could contribute effectively to discussing policies affecting his or her ward. There were strong feelings among Labour councillors about this and towards the end of the period of this study improvements were made in the structure of opportunities available to ward councillors. Three changes in particular improved the position of the councillor: the priority area teams, the area housing management committees and the measures taken to improve the ward councillor's access to information about official proposals for their wards.

The priority area teams began to operate in November 1976 principally as a result of an initiative taken by councillor Jeremy Beecham (then Labour group secretary and chairman of Social Services Committee, later to become leader of the Labour group in May 1977). By June 1978 there were twelve teams. With one exception each team covered a single ward (the twelfth covered parts of three wards). The teams had a quite substantial individual budget – between May 1976 and July 1977 average expenditure per team was £28,395 (Harrop *et al.*, 1978, p. 72). Spending powers directly delegated to the teams were limited. They had to submit proposals to the priority areas subcommittee for approval but this subcommittee only rarely refused to accept recommendations.

The teams only covered fourteen out of twenty-six wards but councillors serving on the teams found their standing and opportunities for influence substantially increased. This was not only because the opportunity to control a budget was offered but also because it gave ward councillors a far greater opportunity to enter into a dialogue with officials who were pretty well beyond reach within the constraints of the normal committee system. As Harrop *et al.* put it, the teams gave councillors 'the power and the opportunity to ask often awkward, sometimes stupid questions' (1978, p. 74).

The second change which enhanced the role of the councillor was the introduction of area housing management committees. This proposal was initiated by the district Labour party. Two pilot committees were set up in October 1977. Later, during 1979, a committee was established in each of the eight management areas. These committees operated without pre-meetings, and this made discussion during meetings much less inhibited than in the main Housing Management Committee. The committees received reports containing recommendations which often had the support of the chairman but because there were no pre-meetings Labour councillors were not bound by group standing orders to support the recommendations. Most of the committees discussed reports relatively uninhibited by party affiliations. Few issues were voted on but, when they were, voting took place across party lines.

Thus Labour councillors found that the introduction of the area committees brought two main advantages. First, it meant that they now had the opportunity to be much better informed about policies affecting their wards. Secondly, they were able to operate within a setting which provided them with new opportunities to have a significant effect on policy decisions. These improved arrangements threw into starker contrast the inadequate machinery of the main committees.

The third change concerned the information available to councillors about their wards. Complaints had long been voiced by many Labour councillors about the failure of council departments to inform them in advance about proposals affecting their wards. The new leader took it upon himself to improve the councillors' access to such information. On his initiative the Management Group of officers considered a report on information for ward councillors in May 1978. Two proposals were adopted. First it was agreed that each committee should circulate copies of reports due to be considered by the next meeting of the committee to the appropriate ward councillors. This system became operative in November 1978 and meant at least that a ward councillor would be informed when a committee was due to consider a report on a matter concerning his or her ward. It remained difficult to secure changes in the proposals but at least ward councillors were now not totally in the dark.

A second measure approved by the Management Group concerned the involvement of ward councillors in work in hand. However, the Management Group only agreed that this issue should be the subject of further discussion between the council's members' services officer and the departments. These discussions took place but, predictably, at the time of writing few, if any, concrete improvements in the councillor's involvement in the planning stages of departmental work had been made.

NOTES: CHAPTER 12

1 In presenting these case studies it has been necessary to refer to myself. Constant use of 'I' can be offensive to some readers and so I refer instead to 'the Councillor'.

2 There were four assistant directors in 1975. The Councillor spoke to the assistant director responsible for administration and finance.

3 This long conversation was recorded by taking notes at the time and then amplifying them immediately afterwards. Wherever possible this was the technique I used throughout. Occasionally note-taking was not possible at the time. In such cases I recorded conversations immediately after they took place.

4 Money allocated under s. 105 of the 1974 Housing Act could not be spent on housing less than thirty years old. The Councillor wanted to know whether this applied to the special sum since much of the housing in his ward was built within thirty years yet needed improving.

5 3 All ER 226.

6 97 LT 431.

7 116 LT 213.

8 A pamphlet produced by the city's Education Department setting out the achievements of the education service between 1974 and 1978 (Newcastle upon Tyne City Council, 1978).

Part Five

THE GROUP AND THE CITIZEN

Chapter 13

THE CITIZEN'S ACCESS TO OFFICIAL INFORMATION AND TO THE COMMITTEE MEETINGS

> In a democracy, people should have information about the workings of government. Without it, they cannot call their representatives to account and make informed use of their rights as citizens and electors.
> (Home Office White Paper, Cmnd 7285, July 1978)
>
> The quality of local government decision-making is not improved by having the spectator breathing down councillors' necks.
> (Sir Harmar Nicholls)[1]

The Redcliffe-Maud Committee argued that one of the two 'essential safeguards for honesty in local government' was 'maximum openness on the part of all those concerned' (1974, p. vii). (See also the Salmon Commission, 1976, p. 69, para. 231.) One of the principal contentions put forward in this book is that the efficiency of local authority decision-making can be substantially improved if it is carried out in the open and in a critical atmosphere. The most significant constraint on the effective working of government along the lines I have described is the limited access to information of the citizen. Indeed, improving the citizen's access to information is possibly the single most important reform which remains to be carried out to improve the health of democracy in this country.

I consider below two aspects of the openness of local government in Newcastle. First, I discuss the degree of access to information enjoyed by members of the public. To what extent were the committee and other official reports made available to citizens? And to what extent were citizens able to obtain information on specific topics, whether or not it was contained in official reports? Secondly, I discuss the extent to which committee meetings were open to the public.

THE CITIZEN AND OFFICIAL INFORMATION: THE
ATTITUDE OF LABOUR COUNCILLORS

Boaden found in Labour-controlled authorities that public attendance
of council meetings was lower, that council documents were less available to the press and that admission to meetings was more restricted
(1971, p. 114).

Labour councillors in Newcastle were asked three questions designed
to establish their attitude towards the availability of official information. Statements were read out and councillors were asked to say
whether they agreed or disagreed with them, and then whether they
agreed or disagreed strongly.

The statement in question 82 was taken from *Labour's Programme for
Britain, 1976* (1976a, p. 104): 'Individual citizens must have a clear right
to information about local government decisions. The burden must be
squarely placed on the council to justify withholding information.'

Thirty-nine Labour councillors answered this question. Of these,
thirty-seven (95 per cent) agreed and two (5 per cent) disagreed. Of the
thirty-seven, there were twenty-nine (74 per cent of the total) who
agreed strongly. What follows are some of the Labour councillors'
responses:

I wouldn't withhold any information from the public. (a subcommittee vice-chairman who agreed with the statement)

They've got a right to it. (a subcommittee vice-chairman)

The above remarks indicate strong support for the statement. However, some councillors who said they strongly agreed with the statement
expressed reservations. For example:

I agree but what about the cost? How many copies would there be? (a
main committee vice-chairman)

Withholding information from the public can be necessary at times.
(a subcommittee chairman who disagreed with the statement)

The statement in question 83 was the policy of the Newcastle district
Labour party: 'Committee reports (other than confidential ones)
should be available to any member of the public before the committee
meeting at the same time as to committee members.' It had been considered by the Labour group in April 1975 but had been rejected.

Forty Labour councillors answered this question. Twenty-seven (67
per cent) agreed and thirteen (32 per cent) disagreed. Of the twenty-
seven, fourteen (35 per cent of the total) did so strongly. This is what

Labour councillors said while responding to this question. Labour councillors who agreed:

> They are available to the press, so why not the public? (a main committee vice-chairman)

> If there was a big demand, yes. The press already do. (a subcommittee chairman)

Some expressed reservations or added riders. For example:

> I agree but they should pay for it. (a main committee chairman)

Labour councillors who disagreed:

> I agree in theory, but in practice? A report could be given to the public and then the committee might disagree with it. They [the public] could be misled into thinking it had been agreed. It would depend on which members of the public. (a main committee chairman)

> The committee is the overriding body. Therefore I disagree. The committee are entitled to expect a certain amount of trust. (a subcommittee chairman)

One backbench councillor was worried that such a practice might – as he put it – cancel out elections:

> Not at the same time as to councillors. I can't see what good it would do. There would be chaos. You would get all the opinions and never get anywhere . . . It would be a killer for politics. If the Labour Party was in power and people and organisations could come back at you it would be a way of cancelling out elections.

Forty Labour councillors responded to the statement in question 68: 'Information should be put on pink paper when the issue under discussion is politically sensitive.'[2]

Twenty-three (57 per cent) agreed with the statement and of these thirteen (32 per cent of the total) strongly agreed. Seventeen councillors (42 per cent) disagreed and only four of these (10 per cent of the total) disagreed strongly. This is what Labour councillors said while answering this question. Labour councillors who agreed:

> We should keep it within the group till the time is right. The public should not know before we decide. The press gives them another

version of what's going on. People can get the wrong impression. It should be given out in the correct way. We may not be able to do what is said in the report. (a backbench councillor)

For example, if we dropped a clanger. It would not be good for the press to know. We shouldn't be so naive as to give our opponents ammunition. We are in a conflict situation, after all. (a subcommittee chairman)

Labour councillors who disagreed:

This is a silly one. You should ignore political sensitivity altogether. (a backbench councillor)

When does an issue become politically sensitive? If we have done something wrong? I disagree. (a main committee vice-chairman)

Thus the attitudes of Labour councillors towards the availability of information were mixed. What degree of access to official information did the citizens of Newcastle enjoy in practice?

INDIVIDUAL REQUESTS FOR INFORMATION: THREE CASES STUDIES

I have already referred above to a common objection raised against case studies of this kind, namely, that they may be discounted as valid findings because the events only took place as a consequence of the behaviour of the participant observer. The argument advanced above in Chapter 12 applies *a fortiori* to the case studies in this chapter. In the examples below the City Treasurer, his deputy and prominent officials in the Administration Department had no reason whatever to behave exceptionally towards a postgraduate student or anyone else.

The first two examples took place in late 1973 and early 1974, when the observer as citizen (before he became a councillor) was trying to gather basic information about the local authority.[3] He had set out to study the accounts of the authority and the voluntary register of interests, the existence of which had been publicised in the local press.

(1) *A Citizen Asks to Inspect Orders for the Payment of Money*

The Citizen quickly discovered that the published accounts did not go into enough detail and, on learning that any citizen had a legal right of access to orders for the payment of money, he set out to get the additional information he sought by exercising this right.

Section 283(2) of the 1933 Local Government Act (in force at the time of the example) lays down a local government elector's right to inspect orders for the payment of money. The section reads: 'A local government elector for the area of a Local Authority may inspect and make a copy of or extract from an order for the payment of money made by the Local Authority.' (The 1933 Act was in force until 1 April 1974; the 1972 Local Government Act was in force after that date and contains exactly the same provision.) Under Section 283(6) the right is exercisable at all reasonable hours without payment. Section 283(7) lays down the penalties which apply if the council obstructs a citizen trying to exercise his or her right. The Citizen rang up the City Treasurer's department and spoke to a senior official. The Citizen explained that he wished to inspect an order for the payment of money as he was entitled to under the 1933 Local Government Act. 'I've never heard of that', was the officer's response. The Citizen explained that he was absolutely certain and asked the officer if he would mind checking up. The officer agreed to check the Act and the Citizen said he would ring back the next day.

When he rang the next day the officer said, 'I've put someone on to this and I can find no reference to a right to inspect orders for the payment of money. You do have a right to inspect the accounts when they are put on deposit under the district audit regulations but I'm afraid that's all.'

The Citizen replied that he was quite sure and read out the relevant section of the Act. The official looked up Section 283(2) and replied that it did appear to lay down a power to inspect orders for the payment of money. He went on to say that he could not do anything further himself without checking with the Deputy City Treasurer. He asked the Citizen to ring back the next day.

The Citizen rang again. The officer said, 'Yes I have spoken to the Deputy City Treasurer. There is no such thing as an order for the payment of money. The nearest we have is either the monthly tabulations [a computer printout of all payments made in a calendar month] or the vouchers [paid invoices or receipts for every payment the local authority has made] but he has decided that you cannot see them.'

'He has decided that I cannot see them. Why is that? There is a clear legal right.'

'There is no such thing as an order for the payment of money; we only have the tabulations or the vouchers.'

'They are the same thing, and there is a clear legal right.'

'Well, anyway, it doesn't matter. Whatever the law is the final say so lies with the City Treasurer.'

The Citizen replied that he would therefore try to see the Deputy City Treasurer. He rang his secretary and arranged an appointment for a few days hence.

At the meeting the Deputy City Treasurer began by asking, 'Why do you want to look at these documents?'

'I just want to exercise my legal right.'

'Yes, but is there any more to it?' he asked.

'I simply wish to exercise my legal right, the purpose is irrelevant.'

The official was not satisfied and examined the Citizen further. 'Are you trying to find out about any particular individual?'

The Citizen replied that he was not. He explained that he was at the local university carrying out research into local government. All he wanted to do was to inspect orders for the payment of money. The conversation continued and the Deputy Treasurer began to make concessions. He explained, as his colleague had done, that there were monthly tabulations. The Citizen had heard that there were two types of tabulation, one which recorded the budget heading (and hence the project or policy the payment was for) under which payments had been made, and one which only recorded the amount of the payment and the payee. The officer conceded this. The Citizen asked if he could inspect the more detailed tabulations but the Deputy Treasurer said he would only let him see the list of cheque payments.

'Does it say what the payments are for?' the Citizen asked.

'No, the tabulations would (he was referring to the more detailed tabulations) but you can't have them. An order for the payment of money does not include details of what project a payment is for, but it would tell you how much was paid and to whom payment was made.'

'But a payment might be made to a particular firm for several different projects. I would not be able to find out how much had been spent in pursuance of some particular policy.'

'No.'

'Well I will have to accept that. When can I see these lists?'

The officer took the Citizen to a room in another corridor and spoke to another official. After he had spoken to him he said, 'You will have to fit in around the availability of space. If there is no space in here then that's unfortunate.' He turned to the other official and asked him whether space was available.

'There will be a desk free on Wednesdays', was the reply (an official was on day-release on Wednesdays). As a result the Citizen agreed to return the following Wednesday. He knew that he had a right of inspection 'at all reasonable hours' but he did not make an issue of this.

The Citizen's efforts were, then, partly successful. He had tried to gain access to information which he hoped would tell him: (i) how much money was paid, (ii) to whom the payment was made and (iii) what the payment was for. He had been allowed to see a list of cheque payments which told him how much was paid and to whom payment was made but not what the payment was for. Sometimes this could be inferred from the amount and the name of the payee but in the main it could not.

The Citizen met with this partial success after a considerable expenditure of time and energy, and after a number of setbacks. His initial request had been met with disbelief. At the second attempt the existence of the statute had been denied. At the third attempt his request was flatly refused. At the fourth request he had been questioned about his reasons for wanting to exercise his right. Eventually his point had only been partially conceded. Most citizens would have been put off at one of these earlier stages, if not because they felt intimidated, then because most citizens would not have had the time to pursue the matter in the way that was necessary.

(2) *Access to the Voluntary Register of Interests*

This example concerns information to which a citizen had no statutory right of access. The local authority had announced publicly, nevertheless, that the information in question was available to the public. Newcastle council had established a voluntary register of councillors' and chief officers' interests, listing shareholdings and other business interests. This was a subject about which there had been considerable national publicity, and great play had been made locally of the establishment of the register. The Citizen rang up the council's Administration Department and asked to inspect the register of interests. He was put on to an officer who confirmed that it would be in order to call at the civic centre to inspect it.

The Citizen went immediately to the civic centre. The officer to whom he had spoken came out to the public inquiry desk and said, 'I am very sorry, I am afraid I have given you false information. The Act in fact says that you cannot inspect it; only a member of the local authority can inspect it.'

'You mean the register is not open to the public at all, only to council members?'

'Yes, I'm very sorry to get you in here for nothing. I know I said it was OK on the phone, but I was speaking off the top of my head – I should have checked first.'

The Citizen was somewhat taken aback and thought that he would have to accept the officer's advice. He had not known that the register was referred to in a statute. Before he left he asked the officer if he objected to telling him which Act he was referring to: 'You mentioned an Act, which Act is that?'

'Er, the 1972 Local Government Act.'

'Could you just tell me which section? Do you mind if we just go along to your room and look it up?'

'No, certainly.'

They walked to the officer's room and he produced the 1972 Local Government Act. 'Which section is it?'

'Section 96.'

The Citizen looked it up and read it whilst the officer sat at his desk. After reading it he said, 'This refers to statutory declarations of interest made in meetings and a register kept in connection with them. That is not what I meant. I was referring to the voluntary register of interests which I read about in the papers. I thought I read that Newcastle had agreed to keep a voluntary register which would be open to the public. The *Evening Chronicle* had a report on it.'

The official appeared as if he did not know how to respond and then said, sounding surprised, 'Oh, you mean the *voluntary* register of interests. Yes, I think that can be seen. But I will have to check.'

He consulted his superior out of earshot in another part of the room. The senior official then left the room and returned with the register and gave it to the Citizen to inspect. Both officers somewhat ostentatiously watched and then the senior officer left the room. The Citizen asked the first officer why he tried to stop him getting the register.

'I've never worked in an authority that was so secretive,' he replied, 'I don't know why there is so much concern with secrecy.'

At this point the senior officer returned and noticed that the Citizen was writing.

'You are not making a copy of it, are you?' said the officer, in a tone of disapproval. The Citizen replied that he was making some notes and that he did not see what was wrong with doing so. If he could see it then there was not much difference between remembering its contents and writing them down.

'We have to obey the law', said the officer.

'But this is not a matter of the law, it's a matter of local discretion.'

'It's the law,' asserted the official. 'We must obey the law. There is no discretion, we just have to carry out the law.'

The Citizen continued to make notes and the senior officer again left the room. A high proportion of citizens would probably have been put off by the first officer's initial response, namely, that it was against the law. In this case the Citizen was able to get his way by being persistent.

(3) *The Lower Heaton Housing Action Area*

This example concerns a request for information, not by an ordinary citizen, but by the observer as a Labour Party local government candidate, who was also chairman of the local residents' association, referred to here as the Chairman.[4] Lower Heaton had been declared a housing action area in March 1975 and a start on a pilot modernisation scheme of three properties promised for July 1975. A letter from the housing action officer (HAO) for the area dated 17 July explained that it had been delayed and that no new starting date would be known until a decision had been taken 'at the end of August'. There was no further reply and so on 19 September the Chairman wrote to the HAO asking when it was proposed to start work on one of the properties in the pilot scheme.

A month later he had received no reply and on 21 October rang up the

HAO and asked him why he had not written back. 'Because of the decision about acknowledgements', was the response. This referred to a recent package of economy measures agreed by the council, which included a decision to stop the practice of sending acknowledgements to letters. The issue was, would it have been an acknowledgement to reply to the letter of 19 September?

The Chairman asked the HAO how a reply to his letter could possibly constitute an acknowledgement. The answer appeared to be that it was an acknowledgement because the HAO did not know when the scheme would begin.

The Chairman probed further: 'Is it because the answer is in the negative that you consider the answer to be an ''acknowledgement'' rather than a ''reply''?'

'Yes', replied the HAO.

The Chairman asked further: 'Do you mean that if there is no affirmative answer to a question but only a negative answer then the latter constitutes an ''acknowledgement'' until such time as there is a positive reply (i.e., you need not reply to my letter, following the economy measures, until there is a date you can name)?'

The HAO replied, 'Yes, in this case'.[5]

The term 'acknowledgement' was being used in a somewhat novel way. A city councillor had suspected that the decision not to acknowledge citizens' letters could lead to abuse and had written to the Chief Executive about it on 10 October 1975. The councillor had suggested that it could result in a far worse service being provided for the public by some departments. The Chief Executive did not agree. He thought that it would lead to the provision of a better and not a worse service. He wrote:

What I effectively wish to put an end to is the tactic of sending out an acknowledgement which so many officers feel leaves them free for three weeks to put off the replies. Instead of hiding behind the reply postcard at 6½p per time, by the thousand, management will endeavour to ensure that a real reply is transmitted as quickly as possible. This operation is not about ignoring the public, but giving them a far better service.

This incident reveals how officials may use almost any rationale, however remote from its intended purpose, to conceal their intentions. The property in question was derelict and had been empty for some months; it had become not only an eyesore but also a health hazard. The HAO was embarrassed by the council's failure to tackle the problem and by refusing to reply was concealing the fact that there was no likelihood of anything being done about it in the near future. That was the residents' suspicion, and was what they were trying to confirm. In fact no work was done on the property until March 1977.

ACCESS TO THE COMMITTEE REPORTS: A CASE STUDY

During the early period of this study there was dissatisfaction among members of the Newcastle district Labour party about the secretiveness of the local council. This dissatisfaction led to a campaign by party activists to widen public access to the council's official committee reports. The district party adopted the following resolution:

> All council minutes and reports and all council committee minutes and reports should be available to anyone on request at the same time as to councillors. This would involve increased expenditure and full cost price should be charged. To predict demand there should be a subscription system.

In April 1975 the Labour group executive committee considered the resolution and recommended the following to the full Labour group:

> That the council minutes and reports and committee minutes and reports other than those on pink paper should be made available to the public on the basis of a subscription system charging full cost price ... It was not, however, thought desirable to make committee reports available to people other than members of the council before consideration by committees since this could lead to pressures being brought to bear on individual members before they had had an opportunity to discuss the reports with their colleagues and involve difficulties over matters of policy. The executive took into account the fact that many committees consult with local residents affected by particular proposals e.g. in the planning field, and consider that proper use of this system provided a reasonable opportunity for the public to be consulted about matters of concern. Extending that concept across the board to policy matters generally was, however, a different matter.

The resolution was a request for regular access to documents which were presented to the council or its committees as a matter of routine. Party members had wanted to have regular access to information and they regarded the routine committee papers as the bare minimum. They had made strenuous efforts to meet their opponents' criticisms. For example, they had met the criticism that the proposed scheme would be prohibitively expensive by recommending that a subscription system be introduced at cost price. Nevertheless, the group executive and the full group were still opposed to the measure. They agreed that reports to the full council meeting should be made available to the public before meetings (they could normally be obtained on request anyway) but did not agree that committee reports should be made available in the same way; they were only to be released after they had been considered by the committee in question. Furthermore, confidential reports were not to be made public at all. (Reports of many kinds were classified as confidential, often without much justification, but they are not the issue here. Some restrictions on access to information can be justified if, for example, it is to protect the privacy of the individual. But in this case the information in question was the ordinary reports to council committee meetings where no such issue was raised.)

Consider the attitude of the group executive committee in the light of Mill's criteria. Did controlling information in this way foster desirable 'moral, intellectual and active' qualities in the population? On the contrary, how could anyone learn anything if relevant information was withheld? To what extent was the machinery of government adapted to take account of any good qualities which existed among members of the public? The machinery was adapted to exclude anyone willing or able to contribute.

However, we do not need to turn to Mill to see the group executive committee's attitude in perspective. At almost the exact time that these decisions were being taken by the Labour group the Department of the Environment and the Welsh Office jointly released a circular on the subject, entitled *Publicity for the Work of Local Authorities* (Circular 45/75). This document stated that great importance was attached to the subject of the circular by the secretaries of state and set out their views:

> It is important that people should be able ... to exert some influence on local decisions affecting their lives: conversely, a local authority must respond as much as possible to the views and needs of those who live within its area. Perhaps even more importantly the public is entitled to know the reasons for the policy decisions of its elected representatives ... In short if the democratic process is to flourish, there must be ready access to full information about a local authority's activities ... *In the view of the Secretaries of State the normal practice should be to let the press have copies of the documents circulated to council or committee members. These should usually reach the press at the same time as they reach members ... There should be no embargo to prevent reports and comments being made in advance of meetings.*
> (pp. 2 – 3, paras 8 and 14, my italics)

The circular does not say that the documents which councillors receive should be sent direct to citizens. But presumably the secretaries of state assumed that the media would inform people of their contents.

It was not only the secretaries of state who wanted to widen access to official information. *Labour's Programme for Britain, 1973* expresses similar sentiments:

Far too often officialdom at all levels shrouds its activities in secrecy
and the public is deliberately denied information about decisions
which can be vitally important to them. We believe that unnecessary
secrecy in government is against the public interest and that
decisions should be taken as openly as possible.
(Labour Party, 1973, p. 89)

The attitude adopted by the Labour group in April 1975 was confirmed later when a report referring to Circular 45/75 was considered
by the council in January 1976. Not only were reports not to be made
available to the public: copies sent to the press were to be subject to an
embargo until noon on the day of the committee meeting. This was in
spite of the advice of Circular 45/75 that documents sent to the press in
advance of meetings should not be embargoed. This move was nevertheless presented as being progressive.

After the publication of Circular 45/75 the Guild of British Newspaper Editors carried out a survey of press and council relations
throughout Britain. It concludes:

Out of either fear or arrogance, many elected members of local
authorities, often with the overt support of their paid officers, are
determined to avoid public discussion of council matters until it is
too late for decisions to be influenced.
(Guild of British Newspaper Editors, 1975, p. 3)

COMMITTEE MEETINGS: THE ATTENDANCE OF THE
PUBLIC AND THE PRESS

Just as cabinet government cannot subsist at all in a public forum so
many other affairs of public life are not helped by being carried on
in public.
(Association of Metropolitan Authorities (AMA), 1975, p. 2)

The AMA then goes on to list some of the principles which in its view
underlie 'the need for appropriate units of informal and private
discussion of business in local government'. These included the
following:

The meeting can properly be used as a forum for the 'floating' of
ideas, and where the admission of the public and press would seriously
inhibit such discussions.

Officers are particularly encouraged to make significant contributions to discussions which, in the political framework of modern
local government, they cannot do at a later stage.

Officers' recommendations may not be accepted and the decision could be the subject of a public inquiry (e.g. planning applications or a court hearing).

The issues are sensitive and members and their advisers might be under undue stress.

Premature disclosure would stimulate unnecessary fears, hopes or activities.

Financial issues of detail are concerned which it would be quite wrong to deal with in public. (AMA, 1975, pp. 2–3)

What is the legal position? What rights, if any, does the public have to attend council meetings? The election of local authorities began with the 1835 Municipal Corporations Act. This Act required that the date, time and place of council meetings should be publicly announced by the exhibition of a notice in a prominent place. It said nothing about any right of the public to attend meetings, and the same continued to be true of the Municipal Corporations Act of 1882, a consolidation of the law half a century later. During these years, however, it became general practice to have benches available for the public at town council meetings. It was usual for their standing orders to provide that the public should be admitted, subject to the right of the authority to exclude strangers and go into secret session by resolution when matters, deemed to be confidential, were to be considered.

In 1908 the Court of Chancery made an important ruling (*Tenby* v. *Mason* [1908]).[6] The Court held that members of the public had no right to attend meetings of the council. As a result Parliament quickly amended the law by passing, in the same year, the Local Authorities (Admission of the Press to Meetings) Act. This remained on the statute book until it was repealed by the Public Bodies (Admission to Meetings) Act of 1960. Subject only to the amendments introduced by the Local Government Act 1972, this is now the substantive law on the matter. Its basic provision is contained in Section 1(1) which states: 'Subject to subsection (2) any meeting of a local authority . . . to which this Act applies, shall be open to the public.' The effect of the Local Government Act 1972 was to widen the class of meetings to which these provisions applied, notably by including committee meetings, but without saying anything about subcommittee meetings. Subsection (2) provides that:

A body may, by resolution, exclude the public from a meeting . . . whenever publicity would be prejudicial to the public interest by reason of the confidential nature of the business to be transacted or for other special reasons stated in the resolution . . . and where such a resolution is passed this Act shall not require the meeting to be open to the public during proceedings to which the resolution applies.

This wording leaves considerable room for manoeuvre. The public interest is a notoriously difficult concept in law (see Boulton, 1978, p. 335); and what does the last part of the section mean by 'special reasons stated in the resolution'?

In the case of *R. v. Liverpool City Council ex parte Liverpool Taxi Fleet Operators' Association* [1975][7] the court held that the requirement of Subsection 2 that special reasons for exclusion must be stated in the resolution had not been met in the particular case. However, Lord Chief Justice Widgery said that 'the requirement that reasons be stated in the resolution' was 'directory only' and the resolution therefore stood 'unless and until set aside by the court'.

In Newcastle the power to exclude the press and public was frequently used. They were excluded for a very wide range of reasons by means of a standard resolution.

A further weakness in the law is that subcommittees and working groups do not have to be open to the public. This can be abused. The Department of the Environment recommended, also in Circular 45/75, that subcommittees be treated in exactly the same way as committees (1975, p. 3, para. 12). However, later in 1975 the Association of Metropolitan Authorities carried out a survey to establish whether or not its member authorities opened subcommittee meetings to the press and public. They report that the position could be summarised as follows:

> Some authorities, but not a large number, open all or most standing sub-committees to public and press and operate the exclusion resolution for the confidential business. But it is clear that much work is done in working parties, panels, steering and liaison groups, joint consultative sub-committees, ad hoc sub-committees ... informal meetings of chairmen and vice-chairmen with officers to deal with detailed matters, and interviewing sub-committees to meet third parties.
>
> Most authorities do not admit public and press to sub-committees but, in some cases, particular sub-committees ... are opened.
> (AMA, 1975, p. 1)

In Newcastle, however, most subcommittee meetings were open to the public, although the press and public were regularly excluded when the subcommittee found it convenient to exclude them.

Although improvements are taking place in some local authorities, they are very much in the minority. The general position in 1977 was such that Professor Harry Street was prompted to write that many of the new large local authorities were 'catching a Whitehall disease: fear of publicity' (1977, p. 567).

NOTES: CHAPTER 13

1   Speaking on the BBC radio programme, *From the Grass Roots*.
2   A report printed on pink paper was regarded as confidential according to council standing orders. Pink reports were considered by committees without the press or public present. The reports were only circulated within the council.
3   I refer to myself as 'the Citizen'. See note 1 to Chapter 12.
4   I refer to myself as 'the Chairman'.
5   Quoted in the Chairman's letter of 21 October and confirmed as an accurate record of the conversation by the HAO in his letter of 28 October.
6   1 Ch. 457.
7   1 WLR 701.

SOME CONCLUSIONS AND POSSIBLE REFORMS

Chapter 14

LIBERTY OF CRITICISM AND THE VALIDITY
OF OUR KNOWLEDGE

It is . . . a maxim of experience that in the multitude of counsellors there is wisdom; and that a man seldom judges right, even in his own concerns, still less in those of the public, when he makes habitual use of no knowledge but his own, or that of some single adviser. (John Stuart Mill, *Considerations on Representative Government*, 1972, p. 333)

In this chapter I return to the themes identified in the introduction. First I return to the problem of modern participatory theories of democracy. I think it can be argued that there is a weakness in the way in which such theories have been supported; a weakness which ought to be attended to if a participatory theory of democracy appropriate to modern conditions is to be formulated. Specifically, too much stress has been laid on ethical arguments for participation and too little on the practical advantages it can bring. This is mainly because 'classical' theories have been interpreted as mainly normative in character. Bachrach, for example, interprets 'classical theory' as being largely ethical, that is, valuing the self-development of the individual, but as being unrealistic under modern conditions: 'Although firmly grounded on what I consider to be a sound ethical position, classical theory falls short of being a viable political theory for modern society' (1969, p. 99). J. S. Mill is referred to in the preceding paragraph as one of the classical theorists Bachrach has in mind. Pateman, who concludes that 'we can still have a modern, viable theory of democracy which retains the notion of participation at its heart' (1970, p. 111) is less prone to this tendency. She recognised that Mill put forward two criteria for assessing the goodness of government. However, she downgrades the importance which Mill attached to the business aspect of government. She writes: 'In J. S. Mill's estimation the merely business aspect of government is the least important; fundamental is government in its second aspect, that of "a great influence acting on the human mind"' (p. 28).

Yet Mill did not advocate greater participation in government merely because it would be beneficial for individual self-improvement; nor was this his *main* concern. Of equal importance was his conviction that it would mean more effective government. In *Considerations on Representative Government* he writes:

> We have ... obtained a foundation for a two fold division of the merit which any set of political institutions can possess. It consists partly of the degree in which they promote the general mental advancement of the community, including under that phrase advancement in intellect, in virtue, and in practical activity and efficiency; and partly of the degree of perfection with which they organise the moral, intellectual, and active worth already existing, so as to operate with the greatest effect on public affairs. A government is to be judged by its action upon men, and by its action upon things. (1972, p. 195. See also p. 11 above)

Some modern theorists of democracy have not only given an unfair emphasis to the views of J. S. Mill. Sometimes they have themselves taken the view that there is a clash between the democratic values which they hold and efficiency values. For example, Bachrach writes: 'The issue for the democrat is not one of choosing between effective government and democratic government. If it were, democracy would be an untenable alternative' (1969, p. 1). A not dissimilar view is held by the sociologist Daniel Bell. He identifies as a source of tension in modern societies the conflict between growing demands for participation on the one hand and increasing professionalism and technical complexity on the other. Speaking about participation he writes: 'the involvement of a greater number of persons *simply means* that it takes more time, and more cost, to reach a decision and to get anything done' (1973, p. 469, my italics).

The view that there is a conflict between democracy and efficiency is widespread. Discussions of democracy in local government are particularly tinged with this notion. A Fabian Society pamphlet states that: 'The ... great choice is between efficiency and democracy ... the demand that there should be vastly greater participation in the machinery of government could lead to total paralysis ...' (Fabian Society, 1971, p. 9). Dilys Hill argues that one essential problem is to 'reconcile efficient services with democratic procedures' (1970, p. 63). Richard Crossman, addressing the annual conference of the Association of Municipal Corporations in 1965 in his capacity as Minister of Housing and Local Government, said that he wanted a new commission to be set up to tackle local government reform. He identified 'the way to resolve the conflict between local democracy, which is inherently small, and efficiency, which is sometimes large' as one of the two problems he wanted the new commission to tackle (Crossman, 1965, p. 660).

Crossman's attitude was shared by the Royal Commission which he set up to investigate local government, the Redcliffe-Maud Commission. The commissioners attached great importance to the size of the population to be provided for by local authorities; they regarded 250,000 as the minimum size necessary 'to command the resources and skilled manpower which they need to provide services with maximum efficiency' (1969, Vol. I, p. 4). A maximum of one million was, they felt, appropriate for the personal services: 'Authorities must not be so large in terms of population that . . . the elected representatives cannot keep in touch with the people affected by their policies' (p. 4; see also pp. 68–73).

The Maud Committee, two years earlier, talked scathingly about English local government democracy:

> The idea that English local government is peculiarly democratic originates in the participation of the members in so much detail. For, unlike the members abroad, they believe, mistakenly in our opinion, that democratic government implies that to discharge their duties they must leave as little as possible to the officers . . . we believe this misconception to be the root cause of local government's administrative troubles, and therefore the reason for our appointment. (1967, Vol. 1, p. 10, para. 40)

They recommended that local government's decision-making machinery should be remodelled along lines similar in some, but not all, respects to central government. The report suggests that councillors should become more like backbench MPs (1967, Vol. 1, p. 125, para. 447). Most councillors were to be members of committees with deliberative but not executive functions (pp. 42–3). Between five and nine leading councillors were to form a management board which would effectively control the authority's decision-making (pp. 41–2). Members of these management boards were not to be 'municipal ministers' (pp. 54–5, especially para. 210). But they were recommended to adopt the practice of accepting collective responsibility for decisions of the majority (p. 55, para. 211b).

Maud's proposals were influential but did not find support in all quarters. The main recommendations of the Bains Report of 1972, on the other hand, met with almost unanimous approval (Greenwood *et al.*, 1974). With the proposals advocated by the Maud Committee, policy-making remained the prerogative of elected members. The Bains Report argued that greater efficiency could best be achieved by shifting the *de facto* if not the *de jure* power to dispose of the resources of the community into the more skilled and appropriate hands of local government officials. The Bains Report quotes with approval the judgement of the firm of management consultants, McKinsey & Co., on an authority they had studied: that 'the democratic forms of council and committee . . . have some great strengths but in many ways are not geared to the modern task of managing thousands of people and hundreds of millions of pounds of assets' (1972, p. 23, para. 4.11). 'We believe', says the report, 'that these words apply with some force to all levels of local authority' (para. 4.12). As a solution, Bains suggested a greater role for the permanent official in the actual formulation of policy (p. 8). Wherever decisions were taken officers '*must* be available' to offer advice; and if the real decisions were taken in the party caucus, then a way '*must be found*' of making the officer's voice heard (pp. 18–19, my italics).

All these official bodies shared the view that local government should be made more efficient. The efficiency they sought was to be achieved by (1) improved personnel management, more training of staff, O & M, work study, manpower planning, (2) the adoption of new management techniques, for example, CBA, PAR, PPBS, PERT, MbO, JAR, OB, operational research, network analysis, critical path analysis and (3) greater co-ordination of policy-making (a euphemism for downgrading the role of the overwhelming majority of elected representatives).

Some of the proponents of the above were seeking cost-effectiveness. In this narrow cost-saving sense their arguments may be reasonable. But invariably advocates of efficiency have made wider epistemological and other assumptions (I return to these assumptions in a moment) and they have usually wanted to achieve their goal by granting more discretion to experts. Some local authority administrators and some local politicians go still further. They have a tendency to regard all discussions of policy by laymen as a waste of time. The following councillors exemplify this sort of thinking.

In an article in *Municipal Review* I suggested some reforms similar to those advocated below (Green, 1977, p. 73) and a number of individuals, active in local government circles, replied to them in subsequent issues. Roland Freeman, a prominent member of the Greater London Council, said that the proposals 'would in practice stop the decision-making process dead in its tracks'.[1] The leader of Leeds Metropolitan District Council (now elevated to the peerage as Lord Bellwin) asserted the primacy of 'making things happen' and implied that rules guaranteeing, for example, wide access to information were undesirable because they would inhibit people (like him) who wanted 'to make things happen' (Bellow, 1977). A third individual, the former leader of Birmingham Metropolitan District Council, drew a distinction between 'discussion, debate and argument' on the one hand and 'decision-taking, action and the implementation of policy' on the other. He concluded by saying 'I cannot emphasise too strongly that members are elected to implement positive policy, not just to talk about it' (Wilkinson, 1977). All these prominent individuals regarded discussion as an obstacle in the way of action. This attitude is even taken by individuals who regard themselves as committed to democracy. John Silkin said, speaking when he was the Labour minister responsible for the new towns, 'To me there is only one great advantage of democracy, and that is that it is accountable. A development corporation can . . . get the right answer without five committee meetings, but it's not accountable.'[2] In saying this the minister reveals a fundamental confusion. But, to his credit, he at least thought that democracy should come before efficiency as he understood it.

What assumptions lie behind the sort of thinking exemplified above? The following assumptions about a single decision-taker (or at least a small number of decision-takers) are made: (1) that the statements of fact which he regards as correct will actually be correct (true) statements; (2) that he can be aware of *all* the relevant information, that is, of all the relevant facts (or statements about what the facts are) and all relevant values; (3) that he will be able to make the right, or best, value judgements. How valid are these assumptions?

In the following pages I attempt to show not only that the above assumptions are unjustified but also that public participation in the consideration of local authority policies is essential if decisions are to result in the best obtainable outcomes. More specifically, I attempt to show that public free criticism is the most effective way – indeed that it is the only reliable way – of arriving at the best obtainable knowledge on any topic.

John Stuart Mill saw very clearly the connection between liberty of thought and discussion and the validity of our factual beliefs. In his essay *On Liberty*, published first in 1859, he wrote:

Complete liberty of contradicting and disproving our opinion is the very condition which justifies us in assuming its truth for purposes of action. (1972, p. 81)

The beliefs which we have most warrant for have no safeguard to rest on, but a standing invitation to the whole world to prove them unfounded. (p. 83)

A similar view was held by the German sociologist Max Weber. In comparing the power of an absolute monarch, in relation to the government bureaucracy, with that of a constitutional monarch, he writes: 'The constitutional king can control . . . experts better because of . . . the *public character of criticism*, whereas the absolute monarch is dependent for information solely upon the bureaucracy' (Gerth and Mills, 1970, p. 234, my italics). This view, held by Mill and Weber, has been put in more sophisticated form by the modern philosopher of science, Karl Popper. I have expounded Popper's view below while considering the validity of the three assumptions identified above.

THE FINALITY OF HUMAN KNOWLEDGE

The first assumption is based upon the belief that certain or secure knowledge can be acquired. Such epistemological optimism can take at least two forms. Either that certain knowledge may be acquired by everyone (or at least by many people) or, on the other hand, that it may only be acquired by a few people (for example by kings, bishops or experts). Epistemological optimism of the first kind can lead to what Karl Popper has called 'the conspiracy theory of ignorance': if a person does not agree with me about the facts then it must be because of malevolence or some selfish interest on his part (1972, p. 3).

The implicit view of commentators typified by Wilkinson, Bellow and Freeman above, is that secure knowledge (in the form of an answer to a question or the solution to a problem) can be acquired but only by a few people (or at least that such knowledge *is already held* by a few individuals). This leads them to regard the involvement of others as a waste of time; they already know the correct version of the facts, no one else could add to what they know, and so for them all that remains is to 'get on with it'. For them democracy is an impediment, it takes too long: 'the best committee is a committee of one.'

How valid is the underlying assumption of these commentators, that there can be some finality about human knowledge? I argue below that there can be no such finality about our knowledge. It is rather, as Karl Popper has persuasively argued, provisional or conjectural. Although Popper argues that there can be no finality about knowledge, he is not an epistemological pessimist and indeed he attacks such pessimism as manifested in Plato. Plato held that abstract entities or universals really existed in an autonomous world of timeless essences. He held that it was only such Ideas or Forms that really existed. Men could have absolutely certain knowledge of them and of nothing else, but in practice their knowledge consisted of imperfect copies of these essences. Popper's criticism is that such pessimism can justify authoritarianism in politics. Although epistemological pessimists take the view that human knowledge cannot be certain but only an imperfect copy of reality it is quite consistent for adherents of this view to argue that the imperfect opinions of a few individuals are very much less imperfect than those of most people and hence that the few should be obeyed (1972, pp. 3–30).

What however, is Popper's criticism of epistemological optimism? He dismisses the optimism of both the seventeenth-century empiricists and the seventeenth-century rationalists or apriorists. These schools can only be understood in their historical context.

In the Middle Ages the determination of truth was typically regarded as a matter for the church and the view held by the church was aprioristic. Although it was definitely aprioristic the Catholic Church was not unanimous in its attitude to truth. It was divided between men who advocated pure rationalism – 'School-Divinity, which in all difficulties useth reason' – and men who asserted that the confirmation of truth should be sought in authority – 'the holy Scripture, Traditions of the Apostles, sacred and approved Councils, and by the Constitutions and Authorities of the holy Fathers' (Whitehead, 1929, p. 11). Whatever these differences, it was certainly held that no individual could, alone, ascertain the truth.

Individualistic criticisms of this view have taken two forms. Religious opponents argued that individuals could discover the truth alone, by revelation. Others argued that individuals could discover the truth alone, by observation in the empirical tradition, and by reason in the rationalist tradition. England was the home of the classical empirical tradition, represented first by Bacon and later by Locke, Berkeley and Hume. They asserted that the ultimate source of all knowledge was observation. They were opposed on the Continent by the metaphysicians Descartes, Spinoza and Leibniz, whose rationalism or apriorism asserted that knowledge, not dependent on experience for its justification, could be obtained by demonstrative reasoning.

What Popper objects to in these theories is that they are potentially sources of authoritarianism. He says that in replacing the authority of the church they have 'appealed to a new authority; the one to the *authority of the senses*, and the other to the *authority of the intellect*' (1972, p. 16). They assume the possibility of an ideal source of knowledge and Popper criticises this view exactly as he criticises the traditional question of political theory, 'who should rule?' The latter question, he says, 'begs for an authoritarian answer such as "the best", or "the wisest", or "the people" or "the majority"'. He argues that it should be replaced by a completely different question: 'How can we organise our political institutions so that bad or incompetent rulers cannot do too much damage?' He criticises the traditional question about the sources of our knowledge – 'What are the best sources of our knowledge?' – in similar terms. This question, he says, assumes that there *is* an ideal source of knowledge. He asserts, on the contrary, that there is no ideal source of knowledge and that '*all* "sources" are liable to lead us into error at times'. He again proposes an alternative question: 'How can we hope to detect and eliminate error?' (p. 25). His theory of knowledge leads him to suggest the following answer to this question: 'By *criticising* the theories or guesses of others and . . . by *criticising* our own theories or guesses' (p. 26).

Consider now how Popper regards human knowledge. A convenient starting point is the problem of induction. Since the time of Francis Bacon the gathering of knowledge has been regarded in this country principally as the task of science. Scientific knowledge has been conventionally regarded (although with much modification) as follows. The scientist carries out experiments. He systematically records his findings and after a time he and other scientists will have accumulated a stock of reliable data. General features emerge and scientists formulate hypotheses seeking to explain how the facts are related to each other. These hypotheses are tested. If an hypothesis is verified then the scientist has discovered a scientific law. This method of basing general statements (whether explanatory or not) on accumulated observations is known as induction.

Hume recognised, however, that no number of confirming instances of any event can justify making an unrestrictedly general statement about it. This realisation has led some thinkers to regard scientific knowledge as no different to any other kind of belief. Fortunately, the problem of induction does not justify such a view, as Popper has pointed out. Popper agrees that no general statement or scientific law can be conclusively verified. For example, consider the generalisation: 'Water boils at 100°C'. No number of recorded observations of it boiling at 100°C could prove that it will always do so. Knowledge is provisional. We cannot prove that any general statement is true and indeed any general statement we make may well be false.

Popper's great contribution was to point out that, although general statements cannot be conclusively verified, they can be conclusively falsified. To return to the example given above: certainly no number of confirming instances of water boiling at 100°C can prove that it will always do so; but any single observation of water boiling at some temperature below 100°C (as it does at the top of a high mountain range) or failing to boil at 100°C (as it does in a pressure cooker) would refute the generalisation. Popper built his methodology on this premise. He rejects the view that all beliefs are equally valid. Nor does he believe that we are unable to justify a preference for one statement over another. We can: by adopting a critical methodology. Briefly, this should involve formulating all theories in a form which exposes them to the greatest possible likelihood of refutation (by maximising their truth content) and then subjecting them to vigorous tests. Because statements can be conclusively falsified Popper argues that these vigorous tests should take the form of systematic attempts to refute the theories rather than attempts to verify them. Popper regards theories which have survived such a process as more reliable than those which have not. (To distinguish between theories tested in this way Popper employs the notion of verisimilitude; see Popper, 1975, pp. 44–60 and 1972, pp. 223–48.)

A crucial point remains to be made. The conjectural nature of knowledge requires not only a critical approach but also that the criticism is a public activity. Popper asserts this because he is very conscious of the fallibility of human beings. For him, everyone is fallible including (perhaps especially) political or official rulers. Though he stresses individual fallibility, this still does not lead him to adopt a relativistic position. He asserts that knowledge has an objective character but that its objectivity does not result from the activity of any single individual but rather from the activity of many individuals. Popper writes:

> If scientific objectivity were founded . . . upon the individual scientist's impartiality or objectivity then we should have to say good-bye to it . . . science and scientific objectivity do not (and cannot) result from the attempts of an individual scientist to be 'objective' but from the *friendly-hostile co-operation of many scientists.* (1966, Vol. 2, p. 217, his italics)

He argues that two aspects of the method of the natural sciences are important in this connection. (He does not regard these methods as applicable only to the natural sciences.) First, the practice of free criticism, that is, of criticising everything from whatever source. Secondly, he says, 'scientists try to avoid talking at cross-purposes'. They do this by 'recognising experience as the impartial arbiter of their controversies'. Popper means experience of a particular kind: 'I have in mind experience of a "public" character like observations and experiments, as opposed to experience in the sense of more "private" aesthetic or religious experience; and an experience is "public" if everybody who takes the trouble can repeat it.' Popper continues, 'In order to avoid speaking at cross-purposes, scientists try to express their theories in such a form that they can be tested i.e. refuted (or else corroborated) by such experience' (p. 218). For Popper there is the world of physical objects (World 1), the world of our conscious thought processes or experiences (World 2), and the world of objective knowledge such as libraries, books, computer memories, etc. (World 3). The knowledge of World 3 is independent of any individual and, most important, it can be improved by many individuals publicly criticising it (1975, pp. 153–190; and 1976, pp. 180–7).

In suggesting this view of knowledge Popper was concerned to put forward a theory that would adequately account for the growth of knowledge. (The way in which Einstein's theories superseded those of Newton particularly impressed upon him the need for this.) He was also concerned to put forward a theory which was oriented towards problem-solving. Government decision-making is predominantly a problem-solving activity and for this reason Popper's insights are of great relevance to modern government. He urges us to 'tackle the practical problems of our time with the help of the theoretical methods which are fundamentally the same in *all* sciences'. He goes on, 'I mean the methods of trial and error, of inventing hypotheses which can be practically tested, and of submitting them to practical tests. *A social technology is needed whose results can be tested by piecemeal social engineering*' (1966, Vol. 2, p. 222, his italics).

Popper's approach and particularly his method of testing the reliability of statements – which the natural and social sciences have in common – dispels the illusion that the best outcomes can be achieved by leaving things to 'the wisest'. More specifically, it dispels the mystique surrounding many of the specialisms to be found in modern official bureaucracies.

To summarise what I have said so far: human knowledge may not be regarded as secure or final. Our knowledge is provisional. This necessitates a critical approach oriented towards the refutation of general statements or theories. This critical approach should have a public character because of the fallibility of human beings. What better argument for a participatory democracy could there be? The very reliability of our knowledge depends upon an activity which must necessarily be public. Any practical and realistic theory of modern democracy must therefore recognise the central importance of guaranteeing the equal right of every citizen to criticise the government effectively.

THE RELEVANCE OF INFORMATION

The second assumption referred to above was that a single decision-taker (or a small group) is likely to be in possession of all the information relevant to solving a particular problem. This will include all relevant values and all relevant facts. Belief in the existence of omniscient fellow men is ill-founded. As Walter Lippman has pointed out:

> Governments are composed of . . . men studying papers at desks, receiving and answering letters and memoranda, listening to advice and giving it, hearing complaints and claims and replying to them; of clerks manipulating more papers . . . These officials have to be fed, and often they overeat. They would often rather go fishing, or make love, or do anything, than shuffle their papers. They have to sleep. They suffer from indigestion and asthma, bile and palpitation, become bored, tired, careless, and have nervous headaches. They know what they have happened to learn, they are aware of what they happen to observe. (1937, p. 25)

It might be argued that the case for the non-existence of omniscient fellow men is not proved by the above comments. It is true that their non-existence is not conclusively proven. Nor, however, is the case for their existence. Indeed there is no evidence at all to suggest that men can become all-seeing and all-knowing, while there is at least some evidence that many men and women are not omniscient. Just as no serious scholars waste time arguing with people who think the world is flat I think it not unreasonable for me to hold the view that no useful purpose would be served by arguing further with proponents of belief in omniscience. I know that some critics disagree with this point of view. But it is legitimate for the social scientist occasionally to have recourse to common sense.

MAKING VALUE JUDGEMENTS

Not only are individuals (or small groups) unlikely to be aware of all relevant values they are also – contrary to the third assumption of my opponents – particularly ill-equipped to make value judgements. There is a tendency to regard the analysis of a policy problem as a process going on wholly in the mind of an individual. Certainly any analysis must involve activity in the minds of individuals but, as I have pointed out in relation to the issue of truth, it is the public character of truth-finding that is essential to success. A similar argument applies to the problem of making value judgements.

It is highly unlikely that value judgements will be best made by individuals or small groups acting alone. If they are made in this way it is likely that the decision-takers will have a view of the facts skewed by their own altruistic or self-regarding values and that they will have ignored the interests of other individuals or groups. Particularly if they are able to disregard the values of other groups or individuals they are also less likely to set out to seek outcomes which will satisfy all interested parties. (It will not always be possible to achieve this; but sometimes it can be done and the success of democratic government will to some extent rest on at least the effort being made and on at least occasional success.)

According to Banfield, the case for having a form of decision-making in which a single expert or team dominates:

> rests upon the assumption that it is possible for a competent and disinterested decision-maker to find in any situation a value premise that uniquely determines the content of the public interest. If there existed several incompatible but equally desirable courses of action a decision-maker would obviously have to employ some 'arbitrary' procedure – e.g. flipping a coin, consulting his own or someone else's personal tastes ... in order to arrive at the decision. But the assumption of administration-minded or planning-minded persons is that this embarrassing situation seldom arises. A competent ... decision-maker ... can usually find in the situation some premise that clearly ought to rule. (1965, pp. 328–9)

'This assumption is wrong', asserts Banfield: 'No matter how competent and well-intentioned, a decision-maker can never make an important decision on grounds that are not in some degree arbitrary or non-logical' (ibid.). He goes on to argue that 'analysis that goes on among many individuals or groups each of whom approaches the problem from his distinctive and limited point of view' may be 'an aid to the correct weighting of values in a choice'. Charles Lindblom's work supports this thesis. He writes:

> Just how does the weighting take place?...Not...in any one analyst's mind, nor in the minds of members of a research team, nor in the mind of any policy-maker or policy-making group. The weighting does not take place until actual policy decisions are made. At that time, the conflicting views of individuals and groups, each of whom have been concerned with a limited set of values, are brought to bear upon policy formulation. Policies are set as a resultant of such conflict, not because some one policy-making individual or group achieves an integration but because the pulling and hauling of various views accomplishes finally some kind of decision, probably different from what any one advocate of the final solution intended and probably different from what any one advocate could comfortably defend by reference to his own limited values. *The weighting or aggregation is a political process.* (Lindblom, 1959, p. 174, quoted in Banfield, 1965, p. 327; my italics)

For Harold Laski the best decisions will be made if the consent of the greatest possible number of persons is obtained:

> The making of policy...is the more successful, the larger the number of affected interests consulted in its construction. The business of government is to draw upon their experience, not as itself it interprets that experience, but as the interests themselves give expression to it. (1967, p. 375)

Where the affected parties to a decision disagree, consent can only be achieved by discussion with them. It may be possible to secure consent for the initial proposal; alternatively, a new proposal may be formulated which meets the objectives of all parties (see Banfield, 1965, p. 253). Where neither is possible, interested parties may be willing to trade gains and losses and so achieve an outcome which, although not intrinsically satisfactory in every respect, has their consent as a whole.

Where none of these options is possible the eventual decision may involve choosing between interests and sacrificing one to the other. Where one interest is favoured at the expense of another, the party which loses out should have the right to be heard. This is a basic principle of justice, even if no concrete result follows from it. In practice someone in this position has an incentive to propose an alternative solution to the substantive problem which may be favoured by all, or at least more, parties. As Lindblom argues, there is a greater incentive to find an agreed outcome where actors are involved in 'partisan mutual adjustment' (Lindblom, 1965, pp. 208–13). Consent is therefore much more likely if all interested parties are able to express freely their views before a decision is taken.

Freedom to express personal preferences is particularly important because so many decisions of modern local government involve making choices between standards of consumption – for example, choosing what internal amenities should be provided in council housing. Consumers should find maximum scope for the expression of their preferences and 'these preferences should be seriously considered as data' (Dennis, 1972, p. 278). Failure to do so may mean that citizens will find their preferences being determined for them by 'professionals . . . who set themselves as the best judges of what each separate family ought to value . . . and what their order of priorities should be' (p. 279).

It is time to summarise what I have said so far. The view that democracy and efficiency are in conflict is mistaken. First, this is because it is not likely that any single individual (or any small group of individuals) will possess at any one time *all* the facts and *all* the values relevant to any particular problem. Secondly, problem-solving necessitates making judgements of value and these are best made jointly rather than individually. Thirdly, no individual or small group will be able, always, to ascertain the facts correctly: knowledge is not definite but rather conjectural. For this reason a critical approach is required. True, all men are fallible, but the likelihood of error can be reduced by adopting a critical approach involving many individuals. Just as the public character of scientific debate strengthens the reliability of its conclusions so would rational free criticism of government – by the 'friendly-hostile co-operation' of many individuals – improve the quality of its policy-making.

The place of public participation in modern theories of democracy can therefore be justified on more than merely moral grounds. Participation in government may well aid the self development of the individual; participatory democracy may well be the only form of democracy consistent with human dignity; but participation conceived as rational free criticism will also increase the chances of government policies producing the best obtainable practical results.

NOTES: CHAPTER 14

1 Letter to *Municipal Review*, no. 572 (August 1977), p. 160.
2 *Labour Weekly*, 23 April 1976.

Chapter 15

SOME PRINCIPLES OF GOOD GOVERNMENT

> Equally lacking is the first condition of all freedom: that all
> officials should be responsible for all their official acts to every
> citizen before the ordinary courts and according to common law.
> (Frederick Engels, in Marx and Engels, 1968, p. 335)

> Let a hundred flowers bloom. Let a hundred schools of thought
> contend.
> (Mao Tse-tung, January 1956)

In this final chapter I consider how the principles expounded in earlier
chapters of this book may be applied in practice in English local government. Before doing this I discuss two objections which are commonly
raised whenever a more participatory form of democracy is advocated,
namely, (1) that there are too many potential participants and (2) that
there are too few potential participants.

THE PROBLEM OF SCALE

Modern societies are too large and complex for direct participation in
very many decisions to be possible. As Macpherson points out, it would
be technically feasible to put in every living room or beside every bed a
computer console with 'Yes/No buttons, or buttons for Agree/Dis-
agree/Don't Know, or for Strongly Approve/Mildly Approve/Don't
Care/Mildly Disapprove/Strongly Disapprove' (1977, p. 95). But at
least two major problems remain: the formulation of the questions and
the intricacy of the answers. It seems inevitable that the government
would have to decide which questions to ask. And, more important, it
would not be possible for the ordinary citizen to respond to many types
of question that would inevitably have to be put to them. For example:

> What increase in the rate of (a) income tax, (b) sales and excise tax, (c)
> other taxes (specify which), would you accept in order to increase by
> blank per cent (fill in [punch in] the blank), the level of (1) old-age

pensions, (2) health services, (3) other social services (specify which), (4) any other benefits (specify which)? (Macpherson, 1977, p. 96)

This is without question a valid problem. Direct participation for everyone in every decision is unattainable. However, direct participation may be possible in some cases where relatively few people are affected. This applies to a number of local authority decisions. For example direct participation – in the form of discussions with officials or councillors – is possible with many applications for planning permission, or when council tenants wish to discuss the details of physical improvements to their homes. With decisions of this provenance citizens will at least be able to have their say. They will still depend on the willingness of policy-makers to take account of their views but the opportunity to express a point of view is still better than no say at all.

However, the problems of scale and complexity do limit the degree to which such participation is attainable. But what is important in examining any particular local authority is whether or not the extent of participation is limited in practice by the problems of scale and complexity or whether it is limited by other factors. In Newcastle upon Tyne the extent of public participation was not actually determined by the difficulties presented by sheer numbers and by complexity. Certainly officers and councillors only had so much time at their disposal but these problems did not determine the amount of direct citizen participation; it was determined rather by the desire of officials and councillors for freedom of action. Generally, the more important the decision the more this applied. It was particularly true of committee decisions. As I have tried to show above, the fundamental weakness of the committee decision-making process was that only a few individuals had an effect on decisions. Once a report which had become the settled view of a chief officer also had the support of the relevant chairman its safe passage was more or less assured; the system principally served to get the reports through rather than to consider them in any effective sense.

Outside the council few – if there were any – individuals or organisations were involved. It was not normal practice to make outsiders aware that particular decisions were about to be taken. On the limited number of occasions when outsiders were told, then it was not the normal practice to make available to them the information available to councillors or officials. Citizens who were able to overcome these obstacles found that opportunities to comment on official proposals were also restricted.

Within the council also, few individuals were involved in taking decisions. Until November 1978 councillors who were not members of the committee taking the decision were not routinely informed that a matter was being actively considered. For those that did know, there were other limitations on their effectiveness. The committee system was heavily reliant on reports. By convention they were prepared by the officers. These reports rarely contained the full facts relevant to an issue. They rarely attempted to discuss alternative ways of dealing with problems. This gave officers a potential to control councillors' perceptions of what aspects of reality were problematic; how councillors perceived such problematic issues; what facts they came to regard as relevant; what propositions they regarded as true or false; what values or preferences they regarded as appropriate. Chief officers and chairmen could also control the timing of the release of official reports. This enabled them to restrict the capacity of councillors to assimilate the relevant information and to prepare a reply.

Furthermore, even in the case of majority party councillors, there were only limited settings in which effective comments could be made – principally the pre-meetings. The extent to which the controlling party (acting as a committee group or the full group) considered proposals was inadequate and usually of a ritual character; caucus meetings tended to legitimise proposals initiated elsewhere rather than to scrutinise them. The role of opposition councillors was negligible.

Often the only elected representatives to give reports any attention at all were the chairmen, sometimes along with their vice-chairmen. And the quality of the attention they gave official proposals varied a great deal.

The position may be summarised as follows. Certainly scale and complexity imposed limits on the degree of direct participation which was attainable. But the actual extent of direct involvement in decisions – by citizens and councillors – was determined not by these problems, but by the desire of leading councillors and officers for freedom of manoeuvre.

LACK OF POPULAR ENTHUSIASM

A familiar argument, advanced by opponents of a more participatory form of government is that very few people want it. Study after study has shown that a large proportion of our citizens are indifferent to politics, that they do not vote and that they understand little about political issues. All this may be true, but the arguments for the kind of participation advocated here are not undermined by this objection. The argument being advanced is that the results of government and specifically the quality of the knowledge utilised by government depend upon public free criticism. Opportunities for citizens to engage in free criticism must therefore be strengthened. If, in the event, there are few takers, then so be it. But the opportunity must be made equally available to all citizens – whatever their social position. The success of government will depend on there being some citizens willing to contribute. And if there are none, well, perhaps all will already have been lost anyway. As John Stuart Mill pointed out, no amount of accountability will matter in a society where the citizens are corrupt:

> Government consists of acts done by human beings; and if the agents, or those who choose the agents, or those to whom the agents are responsible, or the lookers-on whose opinion ought to influence and check all these, are mere masses of ignorance, stupidity, and baleful prejudice, every operation of government will go wrong; while, in proportion as the men rise above this standard, so will the government improve in quality; up to the point of excellence, attainable but nowhere attained, where the officers of government, themselves persons of superior virtue and intellect, are surrounded by the atmosphere of a virtuous and enlightened public opinion. (1972, p. 193).

American experience confirms his view. As Lincoln Steffens reminded his American readers:

> We are a free and sovereign people, we govern ourselves and the government is ours. But that is the point. We are responsible, not our leaders, since we follow them. We let them divert our loyalty from the United States to some 'party'; we let them boss the party and turn our municipal democracies into autocracies and our republican nation into a plutocracy. We cheat our government and we let our leaders loot it, and we let them wheedle and bribe our sovereignty from us. (1904, pp. 11−12)

THE MACHINERY OF GOVERNMENT

However, good government does not only depend on the distribution of good qualities within the population. It also depends on the extent to which the machinery of government allows good qualities in the population to be brought to bear on the activities of government. As modern sociology has taught us, it is the institutions of a society that really count. What citizens lack are mechanisms, which are regularly available for all who want them, (regardless of their resources or the sanctions at their disposal) and which enable them effectively and rationally to criticise the government.

In order to allow popular moral, intellectual and active worth to influence government policies I make the following main proposal: that a set of Principles of Good Government should be (i) enacted by Parliament and (ii) enforced by the ordinary courts of law, as part of the existing process of judicial review.

The arrangements suggested below would alter the relationship between the courts and the local authorities and clearly this raises important issues of principle.

The Constitutional and Legal Background

In the British system there are in practice important overlaps between the legislative, executive and judicial functions of government. Local authorities form part of the executive. They are, like the House of Commons, elected bodies but they do not have legislative powers.[1] The relationship between the executive arm of government and the judiciary rests on two main constitutional principles which combine to produce the doctrine of *ultra vires* – the central tenet of administrative law.

The basic principle is the rule of law. Essentially, this is the doctrine that everything must be done according to law. Every act of a government body which affects the legal rights, duties or liberties of a citizen must be lawful: that is, initially authorised in law and carried out according to legal rules and principles. Anyone who challenges the legality of an act of government may do so before courts which are independent of the executive. The rule of law also means that the government is subject to the law in the same way as an ordinary person (at least to the extent that this is possible, given the nature of governmental functions). The remedies of the ordinary law for trespass, nuisance, negligence, etc., apply to public bodies. In addition there are extra remedies of public law: the prerogative remedies of certiorari, prohibition, mandamus and habeas corpus.

The second underlying principle is the doctrine of parliamentary sovereignty: that Parliament as the legislature is sovereign. The courts have no authority to interfere in any act which is authorised by Parliament. For this reason the courts have sought to justify their decisions by reference to the *ultra vires* principle.[2] This is the doctrine that a public authority may not act outside its powers. There is no such thing as unfettered administrative power. The courts may review any administrative act to determine whether or not it was lawful. If it was not lawful, then a court may quash the decision. This means that the decision is void in law: it is as if nothing had happened. This allows a public body to start again and replace its unlawful decisions with a lawful one.[3]

There is a further important qualification to be made. Although the courts' authority stems from their power to determine whether or not public bodies are acting within the powers granted by Parliament this does not mean that they will only enforce principles which have been expressly approved by Parliament. They enforce a number of other legal principles by means of statutory construction. The existing grounds of judicial review are as follows. (1) Failure to perform a duty: the courts will force an authority to carry out a duty when it has failed to do so; (2) Substantive *ultra vires*: where an Act of Parliament lays down a particular power or duty which is the subject of a legal dispute, the role of the courts is to determine what the statutory language actually means and to apply it to the facts; (3) Procedural *ultra vires*: powers and duties are often conferred by Parliament as long as certain procedures are observed and the courts will enforce these procedures; (4) *Ultra vires* according to other principles such as: (a) exercising discretion unreasonably, (b) delegates cannot delegate, (c) improper use of power, (d) considering irrelevant matters; (5) Breach of the rules of natural justice: *nemo judex in causa sua*, a man may not be a judge in his own cause; and *audi alteram partem*, hear the other side, that a person's defence must be fairly heard; (6) Error on the face of the record: where a court finds an error on the face of the record then different principles apply to those above. It is recognised by the courts that any person or organisation can make a mistake. Because of this an error does not make a decision *ultra vires*. But if an error appears on the face of the record then a court may quash it even though it is *intra vires*.

Judicial Review: the Need for Reform
The view that judicial control of government – including local government – could be improved is widely held. The Law Commission published a working paper in July 1967 to obtain comments on the scope of a possible examination of administrative law (Law Commission, 1969, Appendix A). They reported that 'the large response to our paper and the other evidence of public concern with the problems which it raised clearly favoured the initiation of a study of administrative law as an undertaking of considerable importance and even of some urgency' (p. 1). They set out a summary of the comments they received in an appendix (Appendix C). The Law Commission felt that a Royal Commission should be established to investigate the state of administrative law. The Lord Chancellor, however, took the view that the right time for such an inquiry had not arrived.[4] As a result there was no inquiry.

The Law Commission went on to consider some aspects of administrative law in a later publication. In 1971 they published Working Paper No. 40 on *Remedies in Administrative Law*. This was followed by a report published in March 1976, *Report on Remedies in Administrative Law*. Their main recommendations have now been put into effect.[5]

Although no government inquiry was carried out, the organisation Justice carried out its own. Their report was published in 1971. They found that 'the present arrangements for protecting the citizen' were 'not only unsatisfactory but . . . manifestly capable of improvement' (1971, p. 35). One of their main proposals was that a set of principles of good administration should be enacted. They recommended ten such principles. Some of these are incorporated into my own proposals.

The Courts and the Local Authorities: the Division of Functions
My main proposal is that breach of the Principles of Good Government should be a new ground of judicial review. I have tried to avoid

advocating an arrangement which would involve the courts in taking over functions which are properly regarded as matters for the political process. Certainly, the standard of political activity needs to be improved, but this would not be achieved by handing over functions to the courts. The role of the courts in the proposals advocated below is to enforce a set of principles which are designed to secure a higher quality of political activity.

It is often said that in reviewing the decisions of local authorities the courts are concerned with their 'legality' and not their 'merits'. This division of functions was well described by Lord Justice Roskill in a recent case.[6]

> It is for the council and not for this court to determine what the future policy should be in relation to the number of taxi licences which are to be issued in the City of Liverpool. It is not for this court to consider population growths or falls or the extent of the demand for taxis within or without the city or whether there should be more licences issued in the future than in the past or whether the present grave unemployment position on Merseyside is a relevant consideration. All those are matters for the council. This court is concerned to see that whatever policy the corporation adopts is adopted after due and fair regard to all the conflicting interests.

The distinction between 'legality' and 'merits' is in practice somewhat blurred but it does accurately convey the difference in emphasis between judicial and executive functions. In suggesting a new ground of judicial review – breach of the Principles of Good Government – I have attempted to preserve this emphasis.

Local authority decisions are made up of three elements, law, fact and value. I propose the following division of functions between the courts and local authorities.

Questions of law: Disputes about the correct interpretation or application of the law are for the courts to determine. The Principles of Good Government proposed below would not affect this.

Questions of fact: Findings of fact are traditionally the domain of administrative bodies and not the courts: 'Just as the courts look jealously on decisions by other bodies on matters of law, so they look indulgently on their decisions on matters of fact' (Wade, H. W. R., 1977, p. 274). This is so only as long as the facts are not collateral or jurisdictional, that is, as long as the facts in dispute do not raise questions of the very power of the body to determine the central issue. For example, a rent tribunal has power to fix the rents of furnished dwellings; but if it wrongly regards a property as furnished when it is actually unfurnished[7] or as residential when it is actually used for business purposes,[8] and then goes on to fix a rent, its action will be *ultra vires* and void. In such cases, the facts would affect its jurisdiction; if it found wrongly on the facts the rent tribunal could give itself powers. The courts will always intervene in cases of this type. But as long as an authority keeps within its jurisdiction it may determine questions of fact without fear of intervention by the courts: the facts may be 'found erroneously with impunity' (Wade, H. W. R., 1977, p. 258). There is an exception. The courts will intervene where findings are based on no satisfactory evidence at all. 'No evidence' does not only mean a complete lack of evidence. It includes cases where 'the evidence, taken as a whole, is not reasonably capable of supporting the finding' or, where no authority could 'reasonably reach that conclusion on that evidence' (pp. 274–5). This is how Lord Denning described the rule:

It seems to me that the court can interfere with the Minister's decision if he has acted on no evidence; or if he has come to a decision to which on the evidence he could not reasonably come . . . or if he has taken into consideration matters which he ought not to have taken into account, or vice versa; or has otherwise gone wrong in law.[9]

Should the power of the courts to intervene on issues of fact be extended? Justice recommended that a new right of action in the courts should be introduced, namely, where 'there was a material error in the facts upon which the decision was based' (1971, p. 28, para. 63). They are not the only body to raise this issue in recent years. The Law Commission, in recommending the establishment of a Royal Commission on administrative law in 1969, argued that one of the issues it should explore was whether or not judicial control could be made more effective 'with regard to the factual basis of an administrative decision' (Law Commission, 1969, p. 2).

Since then, signs have appeared that the courts are prepared to extend their control of fact-finding along lines very similar to those proposed by Justice. In a case in 1976 a new ground of judicial review was adopted by Lord Justice Scarman: 'misunderstanding or ignorance of an established and relevant fact'[10] or, as Lord Wilberforce put it in the same case, acting 'upon an incorrect basis of fact'.[11] However, it is too early yet to say how far these developments may go. They are unlikely to make much difference to this issue in the immediate future.

My main concern is to propose measures which will encourage more efficient fact-finding by local authorities in the first instance.

It is important to ensure that political competence is widely dispersed in the community. This would be discouraged if too many aspects of political activity were taken over by specialised institutions like the courts. Furthermore the courts are not always in the best position to determine issues of fact. Such questions can best be determined by the body that actually gathers the facts. For these reasons fact-finding should remain as an activity principally for the political process. The Principles of Good Government which follow are based on this premise, and are designed to improve the quality of local authority fact-finding by seeking to ensure that decisions are based (a) upon correct statements of fact and (b) on consideration of all the relevant facts.

Questions of value: Such issues are very definitely to be determined within the political process. The basic principle adopted here is that issues of value – which may, for example, involve sacrificing one interest to another or asking a group of persons to accept a sacrifice in the public good – should be settled by individuals who are directly accountable to the electorate and, more especially, who can be removed from office at an election. The courts are not accountable in this way. Local authorities are. The role of the courts would be to enforce Principles of Good Government, which were designed to ensure that decisions were not made until all relevant values or interests had found expression.

THE PRINCIPLES

Two objectives underlie the proposals which follow. First, they are intended to improve the speed and quality of local authority decision-making. Principles 1 and 2 are concerned with speed, and the remainder with efficiency. Principles 3, 4 and 5 are based on the premise that efficiency can best be improved by facilitating effective rational criticism of local authorities by citizens.

The second objective was to strengthen the ability of the citizen to judge the local authority by examining its results. It is inevitable that many decisions will continue to be taken by representatives with minimal or no public involvement. The ability of the electorate to judge the performance of government on the basis of its results should not, if the electoral mechanism is to have any significance, be restricted. Many local government decisions and outcomes are visible to the citizenry, but there are many which are not. The utility of the election as a protective device is weakened in direct proportion to the lack of visibility of the actions of local government policy-makers. Greater visibility could be achieved by enacting a Freedom of Information Act modelled on American legislation. This proposal is being widely discussed and nothing more about the need for it can usefully be added here. However, legislation along these lines would not be enough. If passed it would grant the right of access to documents known by citizens to exist. Often, however, a citizen will want to know the facts on which a decision was based and the reasoning lying behind it. This may not be obtainable from known documents. Principle 6 is designed to enable citizens to obtain such information.

Delay

A common complaint about local government decision-making is that there is too much delay. Principles 1 and 2 are designed to prevent unnecessary delay; they are based on two principles proposed by Justice. A distinction has been drawn between decisions relating to a statutory duty and those relating to a statutory power. The principle proposed for decisions relating to a statutory duty is based on experience in the planning field. An applicant for planning permission has a statutory right to receive a decision from the local authority within two months; if no decision is taken the application is deemed to have been refused and the applicant then has a right of appeal to the minister. A simple two-month rule could be applied to all decisions relating to a statutory duty for which there is not already a time limit. The following principle is proposed:

> Where an authority receives a request in writing from any person to make a decision in pursuance of a statutory duty for which no time limit has been prescribed, it shall be the duty of that authority to make the decision to which the request relates within two months of the date of the receipt of the request by the authority. The said period of two months may be extended for a further specific period by agreement between the authority and the said person, whereupon this principle shall have effect as if for the period of two months there were substituted the period as extended. (Principle No. 1)

For decisions in pursuance of a statutory power or discretion the following principle is proposed:

> Where an authority receives a request in writing from any person to make a decision in pursuance of any statutory power, the local authority shall make the decision to which the request relates within a reasonable time from the date of the receipt of the request by the local authority. (Principle No. 2)

Clearly the term 'reasonable' presents difficulties of definition, and will continue to do so until a case law has been developed. There is no satisfactory substitute for the development of such a case law but the following comments may be made. It may be said that for every decision there is an optimum expenditure of time, taking into account (1) the need to act and (2) the need to act wisely. The optimum expenditure of time would be the shortest length of time needed to take the best possible decision; that is, one based on a correct interpretation and application of the law, on consideration of all the relevant facts, on accurate statements of fact and on the most appropriate values. I do not suggest that this optimum time could be strictly delimited. I employ it merely as a regulative notion. But the term 'reasonable' in the above principle could be understood in these terms.

Notice of a Decision, Access to Relevant Knowledge, and the Right to Argue

First, citizens who are affected by proposals should be able to know in advance (i) that a particular decision is about to be taken and (ii) the facts and values (policies or preferences) which the council regards as relevant and on which it will base its decision. Under prevailing conditions information and speed (surprise) may be used as weapons against citizens. The procedural rules proposed below would help to prevent this. Secondly, citizens who are affected by a decision should have a reasonable opportunity to challenge the assumptions of the council and to submit their views before the council takes the decision. These comments, however, need not be made orally. In some cases written comments may suffice.

Individuals in positions of power may well argue that no benefit will be gained by allowing citizens to comment but, as Edmund Burke said in his *Reflections on the Revolution in France*:

> ... in my course I have known, and, according to my measure, have co-operated with great men; and I have never yet seen any plan which has not been mended by the observations of those who were much inferior in understanding to the person who took the lead in the business. (1907, Vol. 4, p. 187)

John Stuart Mill went further than Burke. He felt that:

> ... with respect to his own feelings and circumstances, the most ordinary man or woman has means of knowledge immeasurably surpassing those that can be possessed by anyone else. The interference of society to overrule his judgement and purposes in what only regards himself must be grounded on general presumptions; which may be altogether wrong, and even if right, are as likely as not to be misapplied to individual cases, by persons no better acquainted with the circumstances of such cases than those are who look at them merely from without. (1972, p. 133)

Two principles are proposed. First:

> Before taking any decision a local authority shall give reasonable notice of its intention to take the decision to all persons particularly and materially affected by it; and allow such persons a fair opportunity to make representations to the local authority about the decision. (Principle No. 3)

The formulation 'particularly and materially affected' is taken from Justice. I have adopted it for the reasons advanced by Justice. First, the test of who is likely to be 'materially affected' is objective and therefore limits the possibility of arbitrary interpretation of the phrase. Secondly, by including the term 'particularly' it distinguishes between the public generally and individuals concerned in some additional way. For example, there is a sense in which any ratepayer is materially affected by every local authority decision which involves the expenditure of public money. Thirdly, the formulation avoids the distinction between those who may be harmed by a decision and those who may benefit, a distinction which is inherent in the expression 'aggrieved' commonly used in statutes. At this point, reference should also be made to the changes in court procedure introduced in January 1978.[12] Under the new rules introduced at that time any individual with a 'sufficient interest' was allowed to bring a case before the courts. This is an improvement but it seems to me that the formulation 'particularly and materially affected' is preferable because of the degree of objectivity introduced by the term 'materially'.

Most decisions of local authorities are taken on the basis of official reports. The next principle is designed to give affected persons access to these documents:

> It shall be the duty of a local authority, in proceeding to a decision, to make available to persons who are particularly and materially affected, on receiving a request in writing, any documents which are relevant to the decision except where such documents should reasonably be regarded as confidential. The documents should be made available sufficiently in advance of the decision being taken to allow a person to read them and to prepare an answer. (Principle No. 4)

Procedural rules are especially appropriate because more and more areas of life in this country are being subjected to detailed regulation by the state. As an eminent American judge said:[13] 'Procedural fairness and regularity are of the indispensable essence of liberty. Severe substantive laws can be endured if they are fairly and impartially applied.'

Fact-finding

Two issues are relevant here. First, there is the issue of the accuracy of propositions regarded as correct statements of fact. The principle proposed below requires local authorities to take all reasonable steps to avoid error. Failure to do so would mean that its decision could be quashed in the courts. This still leaves the achievement of factual correctness as largely a matter for the political process. Any errors would still have to be corrected, not by the courts, but within the political or administrative system. The principle would not entitle the courts to intervene merely on the ground that an error had been made, but only on the ground that a local authority had failed to take all reasonable steps to avoid the error.

The second issue is the relevance of particular facts to a decision, and similar considerations apply. A decision would not necessarily be invalid if a local authority merely failed to be aware of a particular relevant fact, but only if the authority had failed to take all reasonable steps to ascertain all the relevant facts. Drawing the attention of decision-takers to the relevance of information would remain a matter largely for the political system.

As Harold Laski said of procedural safeguards relating 'not to the finding of facts, but to the mode of investigating them':

> Unless this type of judicial supremacy is maintained, the executive will always have an overwhelming advantage against the private citizen; and the latter will be unable to invoke his rights against those who shelter themselves beneath the cloak of official acts. (1967, p. 551)

The following principle is therefore proposed:

> It shall be the duty of a local authority, in proceeding to a decision, to take all reasonable steps necessary to ascertain correctly the facts relevant to the decision. (Principle No. 5)

A Duty to Give Both Facts and Reasons

Justice recommended that public authorities should have a duty to give reasons for their decisions. 'No single factor', they say, 'has inhibited the development of English administrative law as seriously as the absence of any general obligation upon public authorities to give reasons for their decisions' (1971, p. 23). The importance which they attached to giving reasons is shared by Lord Denning:

> Every tribunal should give a reasoned decision, just as the ordinary courts do. Herein lies the whole difference between a judicial decision and an arbitrary one. A judicial decision is based on reason and is known to be so because it is supported by reasons. An arbitrary

decision, for ought that appears, may be based on personal feelings, or even on whims, caprice or prejudice. If the tribunals are to command the confidence of the public they must give reasons. (1949, pp. 91–2)

The following principle is recommended:

In relation to any decision it has taken, a local authority shall, upon receiving a request in writing, give a written statement of the facts upon which the decision was based; and of the reasons, including the application of policy, justifying the decision. (Principle No. 6)

A Duty to Provide Accurate Information
Justice also recommended that public bodies should have a general duty to provide accurate information. They say that 'one of the most remarkable gaps in English law is the absence of a remedy for the "irresponsibility" (their emphasis) of the public authority that carelessly provides the individual with inaccurate information' (p. 20). The following principle is proposed to overcome this weakness:

Where a written request is made to a local authority for information relating to the discharge of its duties or the exercise of its powers it shall be the duty of the authority to take all reasonable steps to ensure that such information is provided promptly and accurately, except where it should reasonably be regarded as confidential. (Principle No. 7)

Individuals who regard discussion of the kind which I have advocated above as an unnecessary expenditure of time ignore the inherent complexity of most public issues. Clearly, it is undesirable that time should be spent when its expenditure is not justified by the inherent complexity of an issue. This can happen and should be avoided. Equally clearly, it is undesirable that time should be spent when no benefit at all would be gained by doing so. But how often does such a state of affairs exist? How frequently can we be sure that no advantage at all may accrue from discussion?

Men ignore the real complexity of issues at their peril. This point of view was enunciated, in terms which have never been surpassed, by Pericles in the Athens of 431 BC. He said, while delivering his famous funeral oration in praise of Athenians who had fallen in battle against the authoritarian Spartans:

Instead of looking on discussion as a stumbling-block in the way of action, we think it an indispensable preliminary to any wise action at all. (Thucydides, II, 40; Everyman edn, p. 94)

NOTES: CHAPTER 15

1　Some local authorities have the power under Section 235 of the 1972 Local Government Act to make bye-laws 'for the good rule and government of the whole or any
part of the district or borough, as the case may be, and for the prevention and
suppression of nuisance therein'. However, such bye-laws require confirmation by
the secretary of state.

2　There is an exception. The courts may also quash decisions even if they are *intra
vires* in one type of case, namely, where an error of law appears on the face of the
record.

3　This process of judicial review should not be confused with an appeal. An appeal
means that a superior court or tribunal can *reconsider* the decision of a lower body.
Review is a matter of the legality of a decision: is the authority acting within its
power or jurisdiction (*intra vires*)? If not, a court may declare it legally invalid, but it
may not reconsider it.

4　*Hansard*, HL, vol. 306 (4 December 1969), cols 189–190.

5　By new rules to Order 53, Rules of the Supreme Court, which came into effect on 11
January 1978: Rules of the Supreme Court (Amendment No. 3) 1977. Statutory
Instruments 1977 No. 1955 (L. 30).

6　*R. v. Liverpool Corporation ex parte Liverpool Taxi Fleet Operators' Association*
[1972] 2 QB 299 at 310.

7　*R. v. Blackpool Rent Tribunal ex parte Ashton* [1948] 2 KB 277.

8　*R. v. Hackney (etc.) Rent Tribunal ex parte Keats* [1951] 2 KB 15.

9　*Ashbridge Investments Ltd. v. Minister of Housing and Local Government* [1965] 1
WLR 1320 at 1326.

10　*Secretary of State for Education and Science v. Tameside Metropolitan Borough
Council* [1976] 3 WLR 641 at 656.

11　ibid., at 665; see also *Laker Airways Ltd v. Department of Trade* [1977] 2 WLR 234
at 250.

12　See note 5, above.

13　Jackson, J. (US Supreme Court): *Shaughnessy v. United States* [1953], quoted in
Wade, H. W. R., 1977, p. 393.

Appendix

THE QUESTIONNAIRE

UNIVERSITY OF NEWCASTLE UPON TYNE
DEPARTMENT OF SOCIAL STUDIES
SURVEY OF NEWCASTLE CITY COUNCILLORS

1　Sex.
2　Party.
3　Ward represented.
4　Committee memberships.
5　Chairmanships.
6　Vice-chairmanships.
　　(Questions 1−6 were filled in by the interviewer.)
7　What is your occupation?
8　And would you mind telling me your age?
9　In what year were you first elected on to Newcastle council?
10　Have you been a member of Newcastle council continuously since then?
11　Were you on Newburn or Castle Ward or Gosforth council before that?
12　When were you first elected on to Newburn/Castle Ward/ Gosforth council?
13　Were you a member from then until you became a member of Newcastle council?
14　How many years have you served on Newcastle (and ....) council (altogether) then?
15　Think back to before you were a councillor. What was the first conversation you had with someone about standing for election to the council? Will you recall it for me?
16　In making up your mind whether or not to stand what was the *main* thing that influenced you?
17　Could we talk about what kind of experience being on the council has been for you? What would you say has been the greatest source of satisfaction for you in your council work?
18　What would you say has been the greatest source of dissatisfaction in your council work?
19　When your present term of office is finished do you want to stand for election again?
If 'Yes' ask:
20　How important is this to you? Would you say it was *very important* or *fairly important* or *not important*?
If 'No' to question 19 ask:

21 Would you mind telling me why this is?

22 Now I'd like to ask you how much time your council work takes up. Taking everything into account how many hours did you spend on council work last week? I realise it can only be approximate.

First, how many hours on:

Preparing for meetings ....

Second, meeting with officers (in your capacity as chairman/vice-chairman) ....

Third, meeting with officers (in any other capacity) ....

Fourth, actually attending meetings (including travelling) ......

Dealing with electors' problems ....

Attending other meetings in your capacity as councillor (e.g., as a school governor, or attending residents' association meetings) ....

Anything else ....

23 Total.

24 Was this a fairly typical week?

25 Taking everything into account then, and not just last week, what takes up most of your time as a councillor?

26 Are you able to spend as much time on *all* aspects of your work on the council as you think is needed or are you unable to give some aspects enough attention?

If 'some aspects not enough' ask:

27 Which aspects are you unable to give enough attention to?

28 Now let's turn to your work on committees. The Year Book says that you are on: [names of committees].

Is that right?

29 As a member of the ... [ask of each] Committee you receive reports from the ... [name chief officer]. Do you find it necessary to spend time gathering information *in addition to* that contained in his reports before the meetings?

If 'Yes' ask:

30 As a member of ... [ask of each] Committee do you find it necessary to spend *a lot of time* or *some time* or *only a little time* gathering this extra information?

Now some questions about one of these committees. Let's take the ... Committee.

(For the purpose of questions 31 to 43 a single committee was selected by the interviewer. With one exception, committees of which the respondent was chairman or vice-chairman were not selected.)

31/32 Some people can put a lot of effort into something and get no-where. Whereas, by contrast, other people can get a lot done without having to exert themselves very much. What about the chairman/the chief officer [ask of each]. If he wanted to get the . . . Committee to take a particular decision how much effort, whether before or during the meeting itself, would he have to put in to get his way? Would you say *a lot of effort* or *some effort but not a lot* or would you say *only a little* or *no effort*?

33/34 Now I would like to talk about who has most influence on the decisions which the . . . Committee takes. Would you say that the chairman/the chief officer [ask of each] has a *big effect* on what happens in committee (that is, on the decisions which it takes) or would you say *some effect but not a big effect* or would you say that he has *little effect* or *none at all*?

35/36 Could you say a bit more about this?
(Could you give an example?)
(How does this happen?)

37/38 Still on the same thing. But let's take the individual backbench committee member now. First, the individual backbench committee member in the majority party. (Second, the individual backbench committee member in the opposition parties.)
If he/she wanted to get the . . . Committee to take a particular decision *against the will of the chairman and the chief officer* how much effort would he/she have to put in to get his/her way? Would you say *a lot of effort* or *some effort but not a lot* or *only a little effort* or *no effort* or would it *make no difference* how much effort he/she put in?

39/40 Now still with the . . . [ask of each].
If he/she wanted to get the . . . Committee to take a particular decision *against the will of the chairman and the chief officer* how much effect do you think he/she would have? (How much say do you think he/she would have?) Would you say *a big effect* or would you say *some effect but not a big effect* or would you say *little effect* or *no effect*?

41/42 Could you say a bit more about this?
(Could you give an example?)
(How does this happen?)

43 Still talking about the . . . Committee. May we talk a bit about pre-meetings? How often in your experience are officers' recommendations changed during pre-meetings? Would you say *very often* or would you say *sometimes* or would you say *rarely* or *never*?

Ask chairmen and vice-chairmen:
Could we talk now about your job as chairman of the . . . Committee. I would like to obtain a picture of how things actually work in practice. May I again emphasise that the intention is solely to be objective and accurate so that we

can all have a clearer understanding of the practical processes of government.

44   Does . . . [name chief officer] send you *draft* copies of the minutes of the previous meeting *before they are sent out to other committee members* with the committee papers?

45   Do you ever find it necessary as chairman/vice-chairman to correct the draft minutes so that the version corrected by you goes to committee members and not the chief officer's original draft?

If 'Yes' to question 45 ask:

46   How often would you say you do this? Would you say *often* or would you say *sometimes but not often* or would you say *rarely*? Now could we leave the minutes and turn to committee reports.

47   Does . . . [chief officer] prepare committee reports *in draft* for discussion with you before they go to committee?

If 'Yes' ask:

48   Could you tell me how this works?
(Does he produce two or three copies only especially for discussion with you?)
(Does the discussion take place before the report has been duplicated/printed in sufficient quantity for all the committee members?)
(How long before the committee meeting do you discuss them?)
(Ask chairmen: Is the vice-chairman involved?)

49   Do you find it necessary as chairman/vice-chairman to improve or modify the content of the draft committee reports so that the version as improved or modified by you goes to committee and not that of the chief officer?

If 'Yes' ask:

50   How often would you say you improve or modify these draft committee reports before their submission to committee members? Would you say *often* or would you say *sometimes* or would you say *rarely*?

51   Still on the same thing. How do you see your job when you are dealing with these draft committee reports?
(Do you change the information which they contain or do you modify the policy recommendations or both?)

52   As chairman/vice-chairman of the . . . Committee how much influence over the decisions which the committee takes do you personally have? Would you say *a lot of influence* or would you say *some influence but not a lot* or would you say *a little* or *no influence*?

53   Would you say that you were *satisfied* or that you were *dissatisfied* with the amount of influence which you personally have as chairman/vice-chairman of the . . . Committee? Would you say that you were *very satisfied (dissatisfied)* or would you say that you were *satisfied (dissatisfied) but not very much*?

54 Forget about your position as a chairman/vice-chairman for the moment. Taking only your other work as a councillor (ward work and other committee work when you are only a backbencher): How much say do you personally have when you are a backbencher? (How much influence or power do you have in the decision-making process?) Do you have *a lot of say* or do you have *some say but not a lot* or do you have *a little* or *no say*?

55 Still putting to one side your position as chairman/vice-chairman of the ... Committee, and taking into consideration only those situations where you are a backbencher. Would you say that you were *satisfied* or that you were *dissatisfied* with the amount of influence which you personally have? Would you say that you were *very satisfied (dissatisfied)*?

Ask councillors who are not chairmen or vice-chairmen:

56 Taking into account (committee work, ward work, etc.) how much say do you personally have in the decision-making process? (How much influence or power do you have in the decision-making process?) Do you have *a lot of say* or do you have *some but not a lot of say* or do you have *a little* or *no say*?

57 Would you say that you were *satisfied* or that you were *dissatisfied* with the amount of influence which you personally have? Would you say that you were *very satisfied (dissatisfied)*?
     Let's turn to something entirely different.

58 Think about this. A policy is being considered by a committee of which you are *not a member*. The chief officer's report has been sent out to committee members. You find out about the recommendation. You consider the chief officer's recommendation to be harmful or unjust. Now could you do something about it?

If 'Yes' to question 58 ask:

59 What would you probably do?

60 If you did this how likely is it that you would succeed in getting the chief officer's recommendation changed? Would you say it was *very likely* or would you say it was *fairly likely* or would you say it was *not likely*?

61 Has anyone ever put difficulties in the way of your getting access to information which you needed to carry out your duties as a councillor?

If 'Yes' ask:

62 Has this happened *often* or *on some occasions but not often* or has it happened *only rarely*?

63 Could you give an example? Think back to the last time this happened. Who put difficulties in the way of your getting information on that occasion?

64 Now let's talk about the information which you use as a councillor. Counting everything you do (ward work and committee work etc.),

what is your *main* source of information? Is it:
Your constituents
Your party group leadership
Your local party organisation
Chief officers
Other local government officers
The local press, TV or radio
Tenants' and/or residents' associations
Anyone/anything else

65 What is the *second most important* source of information which
you use as a councillor? ....
May we talk now about confidentiality. By confidential information I mean information which is put on pink paper.
Here are some statements about confidentiality. Would you say
whether you agree or disagree with them? Do you agree (disagree)
strongly or do you just agree (disagree) but not strongly?

66 Information should be put on pink paper when it is necessary to
protect the privacy of the individual.

67 Information should be put on pink paper when it is necessary to
protect the privacy of commercial or industrial organisations.

68 Information should be put on pink paper when the issue under
discussion is politically sensitive.
May we talk now about how you keep in touch with your
electorate?

69 What is the *main* way in which you keep in touch with your constituents?
Now could we turn to something entirely different.

70 Which of the following would you say has the *most power* to bring
about the decisions they prefer? Even when others are initially
opposed to them: [show list]
The leader of the opposition
The leader of the council
The Labour group
The Management Group of officers
The Policy and Resources Committee
The Executive Group of officers

71 Which would you say has the *second most power* to bring about the
decisions they prefer? Even when others are initially opposed to
them.
Now something different again.
Here are some criticisms of chief officers' reports. Would you say
whether you agree or disagree with them?

72 There are occasions when chief officers' reports are too technical.

74 There are occasions when chief officers' reports do not present
alternatives when they should do so.

If he agrees to questions 72/74 ask:

73/75   How often *in your experience* would you say this happens? Would you say *often* or would you say *sometimes but not often* or would you say *only rarely*?

There are some statements here. Would you say whether you agree or whether you disagree with them? Do you agree (disagree) strongly?

76   A councillor should *always* abide by the decision of his party group or pre-meeting in the council or council committee meetings.

77   A councillor should be willing to vote against the policy of his party group or pre-meeting in the council or its committees if he believes it to be harmful or unjust.

78   The councillor should keep his distance from the officers to make sure he will have no difficulty if he ever needs to criticise them or their decisions.

79   The councillor's job is to work closely in partnership with the officers in a close personal and friendly way.

80   All councillors *in the majority party only* should have a right to any information in the possession of the council.

81   All councillors *regardless of party* should have a right to any information in the possession of the council.

82   Individual citizens must have a clear right to information about local government decisions. The burden must be squarely placed on the council to justify withholding information.

83   Committee reports (other than confidential ones) should be available to any member of the public before the committee meeting at the same time as to committee members.

84   Party groups should allow their members to raise in council or in its committees any issue which is not contrary to group policy.

BIBLIOGRAPHY

Association of Metropolitan Authorities (AMA) (1975) *Report to the AMA Policy Committee*, Appendix C (October 1975).

Bachrach, P. and Baratz, M. S. (1962) 'Two faces of power', *American Political Science Review*, vol. 56, pp. 947–52.

Bachrach, P. and Baratz, M. S. (1963) 'Decisions and nondecisions: an analytical framework', *American Political Science Review*, vol. 57, pp. 632–42.

Bachrach, P. (1969) *The Theory of Democratic Elitism: A Critique* (London: University of London Press); first published in 1967.

Bachrach, P. and Baratz, M. S. (1970) *Power and Poverty: Theory and Practice* (New York: Oxford University Press).

Bains Report (1972) *The New Local Authorities: Management and Structure* (London: HMSO).

Banfield, E. C. (1965) *Political Influence* (New York: Macmillan/The Free Press).

Batley, R. (1972) 'An explanation of non-participation in planning', *Policy and Politics*, vol. 1, no. 2, pp. 95–114.

Baxter, R. (1972) 'The working class and Labour politics', *Political Studies*, vol. 20, no. 1, pp. 97–107.

Bealey, F., Blondel, J. and McCann, W. P., (1965) *Constituency Politics: A Study of Newcastle Under Lyme* (London: Faber).

Bell, D. (1973) *The Coming of Post-Industrial Society* (New York: Basic Books).

Bellow, I. (1977) 'Power and the councillor – 3', *Municipal Review*, no. 571 (July 1977), p. 123.

Berelson, B. R., Lazarsfeld, P. F. and McPhee, W. N. (1954) *Voting* (Chicago. University of Chicago Press).

Berry, D. (1970) *The Sociology of Grass Roots Politics. A Study of Party Membership* (London: Macmillan).

Birtles, W. (1978) 'Judicial review of administrative action: the new procedure', *Legal Action Group Bulletin* (August 1978).

Blondel, J. (1958) 'The Conservative Association and the Labour Party in Reading', *Political Studies*, vol. 6, no. 2, pp. 101–19.

Boaden, N. T. (1971) *Urban Policy-Making* (Cambridge: Cambridge University Press).

Boulton, A. Harding (1978) 'Councils, the public and the press', pt 1, *Local Government Chronicle*, 24 March 1978, pp. 318–19; pt 2, ibid., 31 March 1978, pp. 335–6.

Bowring, J. (ed.) (1962) *The Works of Jeremy Bentham*, 11 vols (New York: Russell & Russell).

Brand, J. A. (1973) 'Party organisation and the recruitment of councillors', *British Journal of Political Science*, vol. 3, pp. 473–86.

Brennan, T., Cooney, E. W. and Pollins, H. (1954) 'Party politics and local government in western South Wales', *Political Quarterly*, vol. 25, pp. 76–83.

Briggs, A. (1963) *Victorian Cities* (London: Odhams).

Bulpitt, J. G. (1967) *Party Politics in English Local Government* (London: Longman).

Burke, E. (1907) *The Works of Edmund Burke*, 6 vols (London: Oxford University Press/The World's Classics).

Butterworth, R. (1966) 'Islington Borough Council: some characteristics of single-party rule', *Politics*, vol. 1. no. 1 (May 1966), pp. 21–31.

Buxton, R. J. (1970) *Local Government* (Harmondsworth: Penguin).

Clarke, B., Humphris, A. and James, C. (1977) *The Reasonable Agenda* (London: A. Humphris).

Cockburn, C. (1977) *The Local State: Management of Cities and People* (London: Pluto Press).

Cole, H. B. (1977) *The British Labour Party: A Functioning Participatory Democracy* (Oxford: Pergamon).

Conservative Political Centre (1955) *Rule of Law, a Study by the Inns of Court Conservative and Unionist Society* (London: CPC).

Crossman, R. H. S. (1956) *Socialism and the New Despotism*, Fabian Tract 298 (London: Fabian Society, February).

Crossman, R. H. S. (1965) 'Basic reorganisation of local government', address delivered by Crossman, when he was Minister of Housing and Local Government, to the Association of Municipal Corporations annual conference at Torquay, 22 September 1965, *Municipal Review*, vol. 36, no. 431 (November 1965), pp. 655–60.

Dahl, R. A. (1961) *Who Governs? Democracy and Power in an American City* (New Haven: Yale University Press).

Davies, B. (1979) 'How we aim to increase the councillor's role', *Municipal and Public Services Journal*, 5 January, pp. 11–12.

Davies, J. G. (1972a) 'The local councillor's dilemma', *Official Architecture and Planning* (now *Built Environment*), vol. 35 (February 1972), pp. 112–14.

Davies, J. G. (1972b) *The Evangelistic Bureaucrat* (London: Tavistock).

Davies, J. G. (1977) 'Inside local government', *New Society*, vol. 41, no. 788 (1 September), pp. 438–9.

Dearlove, J. (1973) *The Politics of Policy in Local Government* (Cambridge: Cambridge University Press).

Denning, Sir Alfred (now Lord) (1949) *Freedom under Law* (London: Stevens).

Dennis, N. (1972) *Public Participation and Planners' Blight* (London: Faber).

Dennis, N. (1974) 'Councillors, officers and public participation in urban renewal' in R. Rose (ed.), *The Management of Urban Change* (London: Sage).

Department of the Environment (1975) *Publicity for the Work of Local Authorities*, Circular 45/75 (London: HMSO).

Dicey, A. V. (1905) *Lectures on the Relation Between Law and Public Opinion in England During the Nineteenth Century* (London: Macmillan).

Dicey, A. V. (1959) *Introduction to the Study of the Law of the Constitution*, ed. E. C. S. Wade (London: Macmillan); first published in 1885.

Donnison, D. V. and Plowman, D. E. G. (1954) 'The functions of local Labour parties', *Political Studies*, vol. 2, pp. 154–67.

Donoughue, B. and Jones, G. W. (1973) *Herbert Morrison: Portrait of a Politician* (London: Weidenfeld & Nicolson).

Duncan, G. and Lukes, S. (1963) 'The new democracy', *Political Studies*, vol. 11, pp. 156–77.

Fabian Journal (1952) 'Put policy on the agenda' (February), p. 27 *et seq.*

Fabian Society (1971) 'People, participation and government', *Fabian Research Series, No. 293* (London: Fabian Society, May).

Foot, M. (1975) *Aneurin Bevan, 1897–1945* (St Albans: Paladin).

Forester, T. (1973) 'Anatomy of a local Labour party', *New Statesman*, 5 October 1973, pp. 464–7.

Forester, T. (1976) *The Labour Party and the Working Class* (London: Heinemann).

Freeman, R. (1975) *Becoming a Councillor* (London: Charles Knight).

Freeman, R. (1977a) 'Councillors for courses?', *Local Government Chronicle*, 14 October 1977, p. 825.

Freeman, R. (1977b) 'Cutting down on the committees at county hall', *Local Government Chronicle*, 27 May 1977, pp. 440–1.

Friend, J. K. and Jessop, W. N. (1969) *Local Government and Strategic Choice* (London: Tavistock).

Garner, J. F. (1974) *Administrative Law* (London: Butterworth).

Gerth, H. H. and Mills, C. Wright (1970) *From Max Weber* (London: Routledge & Kegan Paul).

Gibbon, E. (1978) *The Decline and Fall of the Roman Empire*, abridged by D. M. Low (London: Chatto & Windus); first published 1776–88.

Green, D. G. (1977) 'Power and the councillor, a backbencher's right to information', *Municipal Review*, no. 570 (June 1977), pp. 70–3.

Greenwood, R. and Stewart, J. D. (1973) 'Towards a typology of English local authorities', *Political Studies*, vol. 21, no. 1, pp. 64–9.

Greenwood, R., Hinings, C. R. and Ranson, S. (1974) 'Inside the local authorities', in K. Jones (ed.), *The Year Book of Social Policy in Britain 1973* (London: Routledge & Kegan Paul).

Guild of British Newspaper Editors (1975) *Local Government and the Press* (Coventry: GBNE).

Gyford, J. (1976) *Local Politics in Britain* (London: Croom Helm).

Gyford, J. (1978) *Town Planning and The Practice of Politics*, Discussion Paper No. 29 (London: University College, Bartlett School of Architecture and Planning).

Haines, J. (1977) *The Politics of Power* (London: Hodder & Stoughton/ Coronet).

Hampton, W. (1970) *Democracy and Community* (London: Oxford University Press).

Harrop, K. J., Mason, T., Vielba, C. A. and Webster, B. A. (1978) *The Implementation and Development of Area Management*, Area Management Monitoring Project, Second Interim Report (Birmingham: Institute of Local Government Studies).

Hart, J. (1977) 'Freedom and socialism', *Tribune*, 5 August 1977.

Hayter, D. (1977) *The Labour Party: Crisis and Prospects*, Fabian Tract 451 (London: Fabian Society, September).

Heclo, H. H. (1969) 'The councillor's job', *Public Administration*, vol. 47 (Summer), pp. 185–202.

Hewart, Lord Chief Justice (1929) *The New Despotism* (London: Ernest Benn).

Hill, D. M. (1967) 'Leeds', in L. J. Sharpe (ed.), *Voting in Cities* (London: Macmillan), pp. 132–64.

Hill, D. M. (1970) *Participating in Local Affairs* (Harmondsworth: Penguin).

Hill, D. M. (1974) *Democratic Theory and Local Government* (London: Allen & Unwin).

Hobhouse, L. T. (1911) *Liberalism* (London: Williams & Norgate; no date of publication shown, but approximately 1911).

Home Office (1978) *Reform of Section 2 of the Official Secrets Act*, Cmnd 7285 (London: HMSO, July).

Houghton Committee (1976) *Report of the Committee on Financial Aid to Political Parties*, Cmnd 6601 (London: HMSO).

Hunter, Floyd (1953) *Community Power Structure* (Chapel Hill, North Carolina: University of North Carolina Press).

Janosik, E. G. (1968) *Constituency Labour Parties in Britain* (London: Pall Mall Press).

Jennings, R. E. (1977) *Education and Politics: Policy-Making in Local Education Authorities* (London: Batsford).

Jones, G. W. (1969) *Borough Politics* (London: Macmillan).

Jones, G. W. (1973a) 'Political leadership in local government: how Herbert Morrison governed London, 1934–40', *Local Government Studies*, June 1973, pp. 1–11.

Jones, G. W. (1973b) 'The functions and organisation of councillors', *Public Administration*, vol. 51 (Summer), pp. 135–46.

Justice (1969) *The Citizen and his Council: Ombudsmen for Local Government?* (London: Stevens).

Justice (1971) *Administration Under Law* (London: Stevens).

Justice (1978) *Freedom of Information* (London: Justice).

Keith-Lucas, B. and Richards, P. G. (1978) *A History of Local Government in the Twentieth Century* (London: Allen & Unwin).

Kogan, M. and van der Eyken, W. (1973) *County Hall: The Role of the Chief Education Officer* (Harmondsworth: Penguin).

Labour Party (1930) *Report of the Annual Conference* (London: Labour Party).

Labour Party (1955) *Report of the Annual Conference* (London: Labour Party).

Labour Party (1968) 'Report of the Committee of Enquiry into Party Organisation' in *Report of the Annual Conference* (London: Labour Party), pp. 362–80.

Labour Party (1972) *Party Organisation* (London: Labour Party).

Labour Party (1973) *Labour's Programme for Britain, 1973* (London: Labour Party).

Labour Party (1975a) *Conduct of the Labour Party in Local Government, Report of Special Committee of the National Executive Committee* (London: Labour Party).

Labour Party (1975b) *Report of the Annual Conference* (London: Labour Party).

Labour Party (1975c) *Labour Groups on Local Authorities* (London: Labour Party, November).

Labour Party (1976a) *Labour's Programme for Britain, 1976* (London: Labour Party).

Labour Party (1976b) *Guide for the New Councillor and Candidate* (London: Labour Party).

Labour Party (1977a) *Report of the Annual Conference* (London: Labour Party).

Labour Party (1977b) *Local Government Handbook* (London: Labour Party).

Labour Party (1978) *Report of the Annual Conference* (London: Labour Party).

Laski, H. J. (1967) *A Grammar of Politics* (London: Allen & Unwin); first published in 1925.

Law Commission (1969) *Administrative Law*, Law Commission No. 20, Cmnd 4059 (London: HMSO, May).

Law Commission (1971) *Remedies in Administrative Law*, Working Paper No. 40 (London: HMSO, 11 October).

Law Commission (1976) *Report on Remedies in Administrative Law*, Cmnd 6407 (London: HMSO, March).

Lawson, J. (1936) *Peter Lee* (London: Hodder & Stoughton).

Lee, J. M. (1963) *Social Leaders and Public Persons* (London: Oxford University Press).

Lindblom, C. E. (1959) 'The handling of norms in policy analysis', in M. Abramovitz and others, *The Allocation of Economic Resources* (Stanford, California: Stanford University Press), pp. 160–79.

Lindblom, C. E. (1965) *The Intelligence of Democracy: Decision Making through Mutual Adjustment* (New York and London: The Free Press/Collier Macmillan).

Lippman, W. (1937) *The Good Society* (London: Allen & Unwin).

Lipset, S. M. (1960) *Political Man* (London: Heinemann).

Lipson, E. (1947) *The Economic History of England, Vol. 2: The Age of Mercantilism*, 4th edn (London: A. & C. Black).

MacKenzie, W. J. M. (1955) 'Mr McKenzie on the British parties', *Political Studies*, vol. 3, pp. 157–9.

Macpherson, C. B. (1966) *The Real World of Democracy* (New York and Oxford: Oxford University Press).

Macpherson, C. B. (1977) *The Life and Times of Liberal Democracy* (Oxford: Oxford University Press).

Martin, C. and Martin, D. (1977) 'The decline of Labour party membership', *Political Quarterly*, vol. 48, no. 4, pp. 459–71.

Marx, K. and Engels, F. (1968) *Selected Works* (London: Lawrence & Wishart).

Maud Committee (1967) *The Committee on the Management of Local Government* (London: HMSO).
>   Vol. 1 *Report*.
>   Vol. 2 *The Local Government Councillor*.
>   Vol. 3 *The Local Government Elector*.
>   Vol. 4 *Local Government Administration Abroad*.
>   Vol. 5 *Local Government Administration in England and Wales*.

McKenzie, R. T. (1963) *British Political Parties*, 2nd edn (London: Heinemann).

Michels, R. (1968) *Political Parties* (New York: The Free Press); first published in 1911.

Milbrath, L. (1965) *Political Participation* (Chicago: Rand McNally).

Mill, James (1820) *Government* (being an article in the Encyclopedia

Britannica – photocopy held in Newcastle upon Tyne University Library).

Mill, J. S. (1924) *Autobiography* (London: Oxford University Press); first published in 1873.

Mill, J. S. (1972) *Considerations on Representative Government*, published with *On Liberty* and *Utilitarianism* (London: Dent, Everyman Library); first published in 1861.

Mills, C. Wright (1956) *The Power Elite* (London and New York: Oxford University Press).

Mills, C. Wright (1970) *The Sociological Imagination* (Harmondsworth: Penguin); first published in 1959.

Minkin, L. (1978) *The Labour Party Conference* (London: Allen Lane).

Moore, Barrington Jr (1973) *Social Origins of Dictatorship and Democracy* (Harmondsworth: Penguin).

Newton, K. (1976a) 'Who rules the town hall roost', *Municipal Review*, no. 558 (June 1976), pp. 66–7.

Newton, K. (1976b) *Second City Politics* (London: Oxford University Press).

Newton, R. (1968) 'Society and politics in Exeter, 1837–1914', in H. J. Dyos (ed.), *The Study of Urban History* (London: Edward Arnold).

OECD (1978) *Public Expenditure Trends*, OECD Studies in Resource Allocation No. 5 (Paris: OECD, June 1978).

Ostrogorski, M. (1902) *Democracy and the Organisation of Political Parties*, 2 vols (London: Macmillan).

Pateman, C. (1970) *Participation and Democratic Theory* (Cambridge: Cambridge University Press).

Peacock, A. T. and Wiseman, J. (1967) *The Growth of Public Expenditure in the UK* (London: Allen & Unwin).

Peschek, D. (1977) 'Survival kit for councillors', *Local Government Chronicle*, 1 April 1977, pp. 284–5.

Polsby, N. W. (1963) *Community Power and Political Theory* (New Haven, Connecticut, and London: Yale University Press).

Popper, K. (1960) *The Poverty of Historicism* (London: Routledge & Kegan Paul).

Popper, K. (1966) *The Open Society and its Enemies, Vol. 1: The Spell of Plato; Vol. 2: The High Tide of Prophecy, Hegel, Marx and the Aftermath* (London: Routledge & Kegan Paul).

Popper, K. (1972) *Conjectures and Refutations: The Growth of Scientific Knowledge* (London: Routledge & Kegan Paul).

Popper, K. (1975) *Objective Knowledge: An Evolutionary Approach* (London: Oxford University Press).

Popper, K. (1976) *Unended Quest: An Intellectual Autobiography* (London: Fontana/Collins).

Redcliffe-Maud Commission (1969) *Royal Commission on Local Government in England, Vol. 1: Report*, Cmnd 4040 (London: HMSO).

Redcliffe-Maud Committee (1974) *Report of the Prime Minister's Committee on Local Government Rules of Conduct, Vol. 1: Report of the Committee*, Cmnd 5636 (London: HMSO, May).

Redcliffe-Maud, Lord and Wood, B. (1974) *English Local Government Reformed* (London: Oxford University Press).

Redlich, J. and Hirst, F. W. (1903) *Local Government in England*, 2 vols (London: Macmillan).

Rees, A. M. and Smith, T. (1964) *Town Councillors: A Study of Barking* (London: Acton Society Trust).

Ricci, D. (1971) *Community Power and Democratic Theory* (New York: Random House).

Richards, H. (1977) 'Just a minute – members might actually know best', *Municipal and Public Services Journal*, 23 December 1977, pp. 1287–8.

Robinson Committee (1977) *Committee of Inquiry into the System of Remuneration of Members of Local Authorities, Vol. 1: Report, Vol. 2: The Surveys of Councillors and Local Authorities*, Cmnd 7010 (London: HMSO).

Rose, R. (1976) *The Problem of Party Government* (Harmondsworth: Penguin).

Russell, B. (1961) *History of Western Philosophy* (London: Allen & Unwin); first published in 1946.

Russell, B. (1975) *Power* (London: Unwin); first published in 1938.

Russell, B. (1976) *The Impact of Science on Society* (London: Unwin); first published in 1952.

Russell, B. (1977) *Political Ideals* (London: Unwin); first published (USA only) in 1917.

Salmon Commission (1976) *Royal Commission on Standards of Conduct in Public Life: Report*, Cmnd 6524 (London: HMSO).

Schattschneider, E. E. (1960) *The Semisovereign People* (New York: Holt, Rinehart & Winston).

Schumpeter, J. A. (1947) *Capitalism, Socialism and Democracy* (London: Unwin).

Schwartz, B. and Wade, H. W. R. (1972) *Legal Control of Government* (London: Oxford University Press).

Sharp, E. (1962) 'The future of local government', *Public Administration*, vol. 40 (Winter), pp. 375–86.

Sharpe, L. J. (1960) 'The politics of local government in Greater London', *Public Administration*, vol. 38 (Summer), pp. 157–72.

Sharpe, L. J. (1962) 'Elected representatives in local government', *British Journal of Sociology*, vol. 13, pp. 189–209.

Sharpe, L. J. (1970) 'Theories and values of local government', *Political Studies*, vol. 18, pp. 153–74.

Simon, E. D. (1926) *A City Council From Within* (London: Longman, Green).

Smith, B. C. and Stanyer, J. (1976) *Administering Britain* (London: Fontana/Collins).

Smith, S. A. de (1968) *Judicial Review of Administrative Action* (London: Stevens).

Smith, S. A. de (1974) *Constitutional and Administrative Law* (Harmondsworth: Penguin).

Smith, T. D. (1965) 'Local government in Newcastle upon Tyne: the background to some recent developments', *Public Administration*, vol. 43 (Winter), pp. 413–17.

Steffens, L. (1904) *The Shame of the Cities* (London: Heinemann).

Stern, F. (1977) 'Are councillors really necessary?' *Local Government Review*, 1 October, p. 553.

Stewart, J. D. (1974a) *The Responsive Local Authority* (London: Charles Knight).

Stewart, J. D. (1974b) 'The politics of local government reorganisation', in K. Jones (ed.), *The Year Book of Social Policy in Britain, 1973* (London: Routledge & Kegan Paul).

Stewart, J. D. and Boynton, J. K. (1977) *The Councillor's Handbook* (London: Municipal Journal).

Street, H. (1977) 'Secrecy in the town halls', *New Society*, vol. 42, no. 793 (15 December), pp. 567–8.

Thompson, E. P. (1968) *The Making of the English Working Class* (Harmondsworth: Penguin).

Thucydides (1968) *The History of the Peloponnesian War*, trans. Richard Crawley (1876) (London: Dent, Everyman).

Truman, D. B. (1951) *The Governmental Process* (New York: Alfred Knopf).

Turner, J. E. (1978) *Labour's Doorstep Politics in London* (London: Macmillan).

Wade, E. C. S. and Phillips, G. G. (1977) *Constitutional and Administrative Law* (London: Longman).

Wade, H. W. R. (1977) *Administrative Law* (London: Oxford University Press).

Warren, J. H. (1946) *The English Local Government System* (London: Allen & Unwin).

Webb, S. and Webb, B. (1963) *The Development of English Local Government 1689–1835* (London: Oxford University Press); first published in 1922.

Weber, M. (1964) *The Theory of Social and Economic Organisation*, trans. A. M. Henderson and T. Parsons, ed. and intr. T. Parsons (London: The Free Press/Collier Macmillan).

Weber, M. (1968) *Economy and Society, an Outline of Interpretive Sociology*, 3 vols, edited by G. Roth and C. Wittich (New York: Bedminster Press).

Wheatley Commission (1969) *The Royal Commission on Local Government in Scotland, 1966–9*, Cmnd 4150 (London: HMSO).

Whitehead, A. N. (1929) *Science and the Modern World* (The Lowell Lectures, 1925) (Cambridge: Cambridge University Press).

Wilkinson, C. (1977) 'Power and the councillor – 3', *Municipal Review*, no. 571 (July 1977), p. 124.

Wiseman, H. V. (1963a) 'The working of local government in Leeds: Part 1: Party control of council and committees', *Public Administration*, vol. 41 (Spring), pp. 51–69.

Wiseman, H. V. (1963b) 'The working of local government in Leeds: Part 2: More party conventions and practices', *Public Administration*, vol. 41 (Summer), pp. 137–55.

Wiseman, H. V. (1963c) 'The party caucus in local government', *New Society*, vol. 2, no. 57 (31 October), pp. 9–10.

Wiseman, H. V. (1967) *Local Government at Work: A Case Study of a County Borough* (London: Routledge & Kegan Paul).

Wood, B. (1976) *The Process of Local Government Reform 1966–74* (London: Allen & Unwin).

Young, R. (1977) *The Search for Democracy*, (Glasgow: Heatherbank Press).

INDEX

For Product Safety Concerns and Information please contact our EU
representative GPSR@taylorandfrancis.com
Taylor & Francis Verlag GmbH, Kaufingerstraße 24, 80331 München, Germany